THIRD EAR

ALSO BY ELIZABETH ROSNER

NONFICTION

Survivor Café: The Legacy of Trauma and the Labyrinth of Memory

FICTION

Electric City

Blue Nude

The Speed of Light

POETRY

Gravity

THIRD EAR

Reflections on the Art and Science of Listening

ELIZABETH ROSNER

COUNTERPOINT
CALIFORNIA

THIRD EAR

This is a work of nonfiction. However, some names and identifying details of individuals have been changed to protect their privacy, correspondence has been shortened for clarity, and dialogue has been reconstructed from memory.

Library of Congress Cataloging-in-Publication Data
Names: Rosner, Elizabeth, author.
Title: Third ear : reflections on the art and science of listening / Elizabeth Rosner.
Description: First Counterpoint edition. | California : Counterpoint, 2024. | Includes bibliographical references.
Identifiers: LCCN 2024010163 | ISBN 9781640095519 (hardcover) | ISBN 9781640095526 (ebook)
Subjects: LCSH: Listening. | Listening—Social aspects.
Classification: LCC BF323.L5 R67 | DDC 153.6/8—dc23/eng/20240530
LC record available at https://lccn.loc.gov/2024010163

Jacket design by Nicole Caputo
Jacket art © iStock/Homunkulus28
Book design by Olenka Burgess

COUNTERPOINT
Los Angeles and San Francisco, CA
www.counterpointpress.com

Printed in the United States of America

1 3 5 7 9 10 8 6 4 2

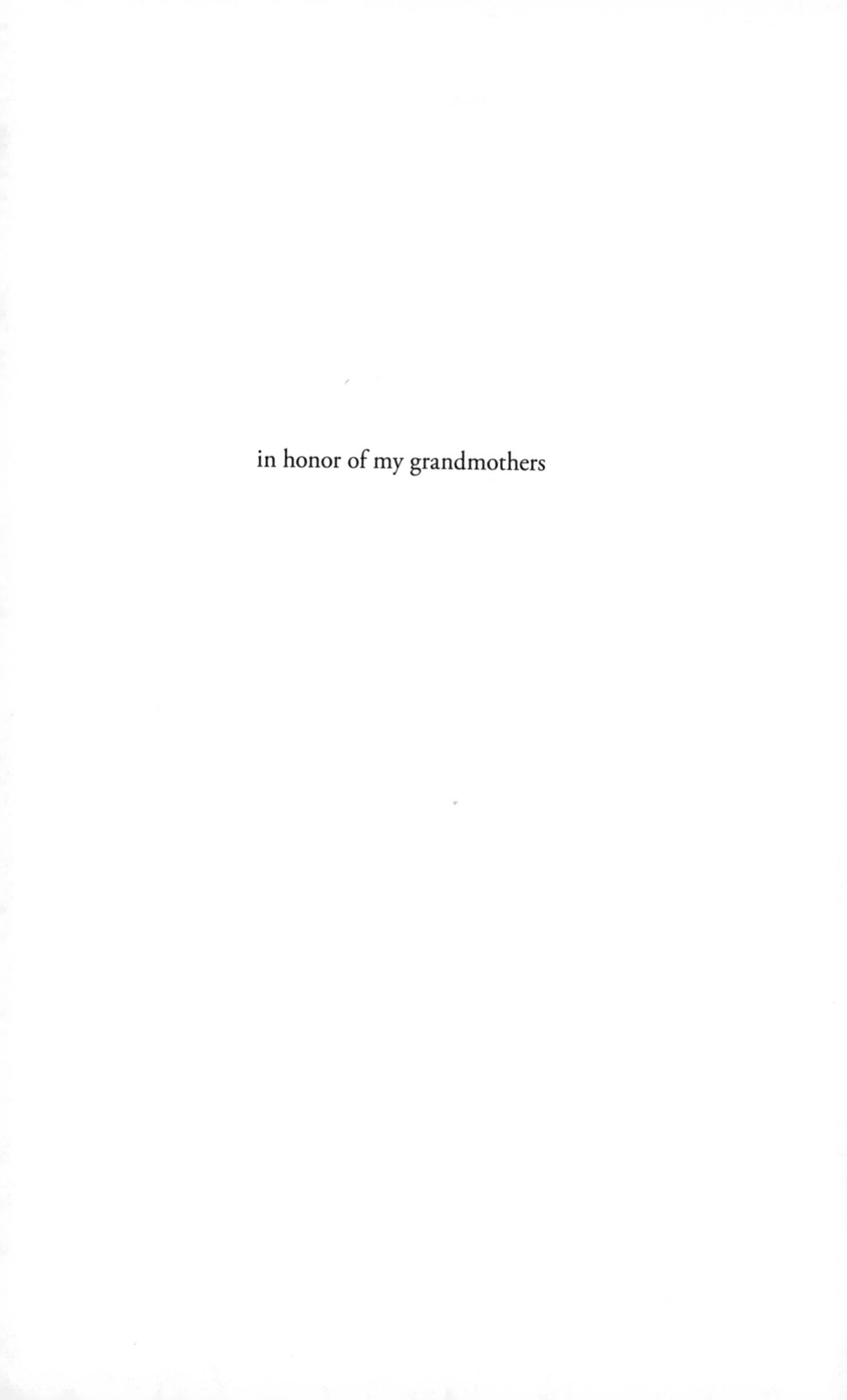

in honor of my grandmothers

Take a walk at night.
Walk so silently that the bottoms of your feet become ears.

—PAULINE OLIVEROS, *Sonic Meditations*

CONTENTS

THIRD EAR

PORTAL, AN INTRODUCTION

YOU COULD SAY THAT HEARING IS A SCIENCE AND LISTENING is an art. Hearing depends on signals received by a functioning apparatus; hearing is measurable and verifiable. But listening is so much more than gathering information with the paired portals of our ears, those supple appendages flanking our faces. We also absorb with heart and skin, our elaborately nuanced nervous system continuously networking.

For an imperfect comparison with the filtering, processing, and discriminating involved through your other senses, consider the distinctions between looking and examining; touching and palpating; tasting and savoring a flavor; smelling and identifying a scent. When we are truly listening, signals are not merely accepted but are fluently interpreted. Transformed into meaning.

Dialogue is happening all around us: over our heads, under our radar, beyond the horizon. Whether or not we decide to call it language, we can call it listening—when elephants respond to news from their miles-distant family members, details they've taken in through acoustic sensitivities in their feet. When the searching roots of trees grow toward the energetic flow of water, sometimes tangling themselves in underground pipes—isn't that too a kind of listening?

What about when hummingbirds return to the specific vibration of nectar-drenched flowers? When whales share songs across oceans and recognize the pauses made by one another's breathing?

Hushed or amplified, implausible yet audible, *everything* is humming—from quantum to cosmic, from the inner life of electrons to the membranes of outer space. The entire universe is sonic.

. . .

Theodor Reik, protégé and colleague of Sigmund Freud, developed a method he called "third-ear listening." He credited Friedrich Nietzsche with the origins of the phrase, although Nietzsche was referring to ways of listening to music.

"The analyst, like his patient," wrote Reik, "knows things without knowing that he knows them. The voice that speaks in him, speaks low, but he who listens with a third ear hears also what is expressed almost noiselessly, what is said *pianissimo*. There are instances in which things a person has said in psychoanalysis are consciously not even heard by the analyst, but nonetheless understood and interpreted. There are others about which one can say: in one ear, out the other, and in the third."[1]

. . .

In at least one Indigenous culture, comprehending with a quiet, respectful awareness known as *dadirri* (an Aboriginal word for deep listening)[2] is a way of being that has been practiced for more than sixty thousand years. Like third-ear listening, *dadirri* focuses with patience and stillness both externally and internally. "One of the peculiarities of this third ear is that it works in two ways," Reik explained. "It can catch what other people do not say, but only feel and think; and it can also be turned inward. It can hear voices from within the self that are otherwise not audible because they are drowned out by the noise of our conscious thought processes."

. . .

As the daughter of multilingual parents, I learned early to attune myself to their sounds and silences, even when they weren't necessarily doing the same for each other or for me. Within their whispers, I detected spaces between clamor and consolation, between foreign and familiar. Maybe this prepared me for a life of eavesdropping on the world, listening with *all* of my senses, reaching toward sources of interconnection.

Reflecting on some of my most profoundly transformative experiences, I note that they have centered on the flowing back and forth of sound—whether occurring in childhood or adolescence, in leaving home or returning home. In a lifelong search for sanctuary and awareness, deep listening has become my way of leaning closer, opening wider, taking more responsibility for honoring the music of everything.

. . .

Every time the resonant world enters us, we have an opportunity to reaffirm our relatedness. Today, for instance, a murmuring podcast entices me with the latest studies of complicated, gorgeous sounds too high and too low for my meager human range. The more I learn about acoustic windows in the feet of elephants and in the skulls of whales, I imagine my body reverberating in cellular empathy. These compositions and collaborations taking place among birds, insects, and cetaceans remind me that I belong to a sonically interwoven web. That includes sounds still echoing toward us from the distant past and sounds we are attempting to decipher in order to cocreate a future. It includes frequencies connecting galaxies and microbes, connecting the dead with the living, connecting the beings we might yet comprehend with the beings yet-to-be-born.

I hope that this book can serve as another kind of portal, with a sequence of spaces shaped by its author's curiosity, questions answered and also sometimes unanswerable. In these pages you will find references to mammals on land, mammals who returned to the depths of the oceans, songs of the air, the forest, the soil. I believe that third-ear listening can offer solace for my/our existential ache of loneliness. With our ancestrally related bones offering and receiving the news, we can lengthen and strengthen the golden thread stitching together a broken world.

FIRST SOUNDS

MY EARLIEST MEMORY OF LISTENING: I WAS ABOUT three, pretending to be asleep in the green canvas hammock stretched above the slab of concrete that I had recently learned was called a patio. In the drowsy quiet of this summer afternoon, my father and mother were murmuring to each other in phrases I wasn't supposed to overhear. "*Hon air soot*," something like that. Even though I didn't know what the words meant (and would not discover for years), I somehow understood that these were private sounds, shared only between the two of them. They were talking about me, but I was not allowed to know my parents thought I was cute, sweet, adorable, adored. Such secrets were for my benefit, to protect me from the harm inevitably poised to strike me down, the way they had both been struck. It was dangerous to speak praise aloud, dangerous to hear it. They whispered all kinds of mysteries to each other, and I was not to know them. But I wanted to.

. . .

Listening begins in utero. Anne Karpf, who—like me—is a daughter of Holocaust survivors, writes in *The Human Voice* that in fetal development our ears are beginning to form as early as two months. By five months, our hearing structure is anatomically complete.[1] The amniotic sea of sound in which I floated would have held an array of my mother's languages: the Russian she used with her own mother; Polish and Hebrew with her best friends; English with my sister and with the American neighbors; Swedish with my father. Even before I was born, my world was a symphonic mother tongue.

. . .

Elephants, growing in utero for nearly two years, are listening and learning too. With the soles of their feet pressed against the throbbing pulse of the womb's lining, they are becoming familiar with the rumblings of their mother and their matriarchal relatives. Pitched at infrasonic levels inaccessible to our ears, these deep vibrations will be the same sounds that greet the newborn elephants upon their release from a watery interior world. Prodded upright by the gentle nudging of mother's trunk, audibly instructed not only to stand but also to start walking (*Let's go, Follow me, Stay close, I'm here*), they begin their new life on land.

. . .

Humans practice empathy and understanding as early as twenty minutes after birth, not just with sound but with our body movements and what has been termed "interactional (or behavioral) synchrony." Those voices we heard in utero remain for quite a while in our physical memory, rhythmically threaded into the flexible reach of our limbs. We speak before we speak, nodding and twisting in the vocal dance of our mothers and our cultures.[2]

Born hungry to be heard, we are wired to make the very sounds we are designed to receive. Just as we push sound waves out of our bodies to vocalize, our eardrums vibrate in response to those waves, resonating at the same frequencies.[3] Infants express a range as nuanced as a mother's ability to decipher the code and to respond accordingly. Using volume and pitch, we signal our needs for food, warmth, contact, and relief.

. . .

In an essay about lullabies and language, Kristin Wong writes about the seemingly universal practice of singing to infants, who are described as "impressively sophisticated listeners."[4] It's as though certain patterns have been inscribed in us, encoded and memorized. "Brains of babies who heard a specific melody just before birth reacted more strongly to the tune immediately after they were born and at four months."[5]

Sally Goddard Blythe, director of the Institute for Neuro-Physiological Psychology, is an expert in early child development. Noting studies of the effects of listening in utero, Blythe explains that the mother's voice "is particularly powerful because it resonates internally and externally, her body acting as the sounding board."[6] Research keeps demonstrating the depth and continuity of this bond, as "a mother's voice provides a connection between respiration, sound and movement, an acoustic link from life and communication before birth—to the brave new world after birth."[7]

. . .

As Anne Karpf explains, attunement and reciprocity are two key features of early childhood development that involve listening skills in *both* parent and child. The rhythmic patterns of preverbal vocal exchanges between babies and mothers are "remarkably similar to the timing of verbal dialogues between adults—not for nothing are they known as 'proto-conversations.'"[8] From our first phases of feeling listened to, nonverbal signals are provided by a silent acknowledgment from the receiver of one's words and stories.

> Nonverbal factors such as nodding, eye movements and facial expressions are used to indicate that a reciprocal relationship between speaking and listening is being played out. As the listener "rewards" the speaker with non-verbal feedback, "a societal level ritual and regulation" is established. The speaker is encouraged to continue, confident in the belief that his message is being received and understood. Even when the facial expressions are perceived by the speaker as being negative or confused, there is still the silent agreement that the ritual of dialogue is taking place as the speaker receives the unspoken message "I'm listening."[9]

. . .

A variety of species, including female greater sac-winged bats and adult male zebra finches, have been recorded using so-called baby talk. A recent report published in the *Proceedings of the National Academy of Sciences* journal reveals that dolphins, much like humans, use a 'baby voice' to communicate with their young, compared to the pitch they use while swimming alone or with adult dolphins in the rest of their pod. After recording the distinctive sounds of mother bottlenose dolphins in the wild in Sarasota Bay, Florida, researchers found that "all 19 of the mothers changed their tone when they used their signature whistle to speak to their baby."

A coauthor of the study, Laela Sayigh, a Woods Hole Oceanographic Institution marine biologist, explained that the signature whistle is understood as a way that dolphins keep track of one another's whereabouts, "periodically saying 'I'm here, I'm here.'"[10]

. . .

Babies hear an even wider range of frequencies than adults, especially in the high-frequency ranges.[11] It turns out that humans across a wide spectrum of cultures have, accordingly, evolved a universal vocal pattern of interacting with newborns using "exaggerated, simplistic sentences in high-pitched, sing-songy tones," writes Kristin Wong.

> Despite how much it makes adults cringe, *Parentese* is good for babies. One study found that it helps them learn new language skills and develop a larger vocabulary. Parentese also makes babies feel safe. When they hear baby talk, babies babble back, which some research even suggests is a way of processing stress. And that's important when you're pulled out of the womb and thrown into a world

> that's unpredictable and blurry. You need someone to translate adult human speech to baby talk. You want someone to tell you it's all going to be okay. To say things like, 'who's got the chubbiest thighs?' to distract from the fear of not knowing when your next meal will be.[12]

Wong writes about her own experiences of learning this new baby language as a mother—and also about learning to hear the words of others, even strangers, with newly sensitive ears. As if all listening is a form of translation.

. . .

We are born with the ability to hear the sounds of all languages, says Dr. Viorica Marian, professor at Northwestern University, whose research specializes in neurological studies of bilingualism and multilingualism. She is referring to something that researchers call "a window of universal sound processing," a built-in capacity to listen that facilitates our earliest learning about and connecting with our primary caregivers. And yet, Dr. Marian says, by the time we reach the age of one, our brains are tuned in to our native language. She points out that this doesn't mean we are born tabula rasa, but more accurately it means we are born without nationality.[13] Studies have shown that this window stays open longer for multilinguals than for monolinguals.[14]

. . .

One Sunday afternoon when I was six, my mother and father were sitting on faded lawn chairs in the backyard, discussing the overgrown vegetable garden they had inherited from our house's previous owner. Pointing into an open space, my mother said she wanted to plant an apple tree, or maybe a pear.

"For baking," she explained.

My father said, "Trees take a long time to grow," adding that even the maples we planted last year in the front yard were barely taller than the children, who helped dig holes for their roots.

Sharon, a friend from the neighborhood, was with me. When we asked permission to go for a bike ride, my parents smiled and waved, told us to have a good time. The year was 1966; our town had streets safe enough and wide enough for us to practice riding with no hands. We could go anywhere we want.

"Your parents talk funny," my friend shouted into the wind as we took off. "They have accents."

I didn't respond, pretending to assimilate her announcement like it was a fact. But in that moment, some design in my awareness began to rearrange itself. I knew Sharon's father worked for General Electric, the same company my father worked for; her mother wore dresses not unlike the ones my mother wore. She had a brother and a sister just as I did. Our houses on the street looked nearly identical, except for the color of the front door and the type of curtains hanging inside the windows. But now it occurred to me that my friend and I didn't inhabit the same country.

That evening, sitting at our Formica dinner table, I heard in a new way how my father pronounced the number three as "*sree*," how my mother said, "You'll pass with *swinging* colors." I'd been a fish in water, not noticing it wasn't air. To someone else, these two voices sounded entirely different than they sounded to me. To "talk funny" and to "have accents" meant *stranger*, meant *foreign*, meant *not like us*. After a while, I would wonder if what Sharon really meant was *not American*.

I had crossed a dividing line in my life. Listening for accents everywhere, my ears were newly tuned to catch some exceptional lilt or word coming from deep in the throat, a signal from places I might have known, but didn't. I began focusing on my mother's intonations during phone calls and over coffee with friends in town whose tongues were shaped in Poland, Russia, Czechoslovakia, Israel. Eventually I found out that Russian was her favorite; it was the one in which she wrote poetry. German, my father's native tongue, was "the language of the murderers." Swedish was the language of exile, refuge, and love—*hon är söt*. It was the one she shared with my father, for secrets.

I shared secrets of my own—with dolls, with animals, with the night sky. For a few years we kept a canary in a cage in the kitchen, though retroactively, protectively, I want to believe we had not one but a pair of them. I can't bear the thought that we would have imprisoned a bird alone—singing in perpetual search for an elusive mate. My mother, who had been an only child, frequently described her solitude; she always wished for a sibling, as if that would have magically eased the ache.

I had an older sister and a younger brother, yet a mood of isolation pervades every single memory from my childhood. I am shadowed by a sense of being unaccompanied in my deepest central self.

. . .

My family legacy of loneliness—and its partial cure—began in Stockholm, where my parents met as young refugees after the end of World War II.

Their map of hope had been outlined a decade earlier, with the journey of my Uncle Eli, previously Helmut, sent as a four-year-old to Sweden via *Kindertransport* from Hamburg around 1938. After desperately placing her other two sons—my father and his other brother—in a Jewish orphanage, my grandmother Rachel smuggled herself to Stockholm to be closer to her youngest, with dashed dreams of getting the other boys out of Nazi Germany before it was too late. Miraculously, my father and his brother Wolfgang survived deportation and imprisonment in Buchenwald concentration camp as teenagers in 1944. Once the camp was liberated in 1945, they made their way to Stockholm to be reunited with Rachel and Eli. (Imagine all of this transpiring across years of uncertainty, across bombed landscapes and numerous episodes of starvation, illness, near-death.)

By this time, my Polish-born mother, also a teenager, was in Stockholm awaiting an American visa while in transit with her parents; the trio had managed to survive being trapped in the Vilna ghetto for two years and, subsequently, more than a year of hiding in the Polish countryside. Although Russians liberated Poland from the Nazis in 1944, continued antisemitism persuaded my maternal grandparents to become refugees in search of sanctuary.

And so, drawn together by some magnetic field of shared trauma and matching optimism, at ages nineteen and twenty, my future parents fell in love and got married in Israel in 1950 (where most of my father's extended family from Germany had fled in 1933). In 1952, my newlywed parents made their way to the United States of America.

No accident that my older sister, younger brother, and I were all born in this adopted country, the nation that had literally saved my father's life. (Buchenwald prisoners were liberated by Patton's Third Army on April 11, 1945, exactly one week past my father's sixteenth birthday.) Citizenship was one way of claiming a new home country, but so was choosing to raise a family in *its* language. The more I heard of my parents' histories, the more I understood that the vowels and syllables of America were, especially for my father, the vocabulary of freedom. English in your mouth and in your ears meant opportunity, ambition, achievement. Most of all, it meant safety.

. . .

Were we safe? My mother used to say—as if recalling a funny story—that as a toddler I cried so loudly and relentlessly she sometimes left the house to escape the sound. "And then you followed me down the street," she would add, "crying."

Every time I think of this, I'm stunned to picture myself wobbling on a pair of chubby little legs, using all my lung power to aim my voice at her back. Only later did I learn about the ways my mother in her own early childhood felt abandoned by *her* mother. Even more excruciating, she told us about the time her young cousin's crying threatened to expose the family in their hiding place during the Vilna ghetto's liquidation. Although she was never quite able to untangle it for me—or for herself—I suspect the intensity of my sobs must have alarmed my mother to the breaking point.

By the time I was a young adult filled with regret for missing out on the chance to learn my mother's beloved Russian, I asked her why she didn't teach me. "I tried singing to you when you were small," she said, "while I bathed you and tucked you into bed at night. The same melodies my mother sang to me. But you held your hands over your ears and chanted, 'English, English.' You refused to listen."

. . .

During our phone interview, when I ask Dr. Marian about her own multilingualism, she explains rather casually: "I was an early bilingual. Romanian was the language we spoke at home, but Russian was the official language of the country. I grew up speaking both of them, so I don't have an accent in either." She later acquired a few more languages, including English as her third.[15] I feel insatiably curious about the changing shape and dimension of my own inner portals of listening, the ones formed in utero, my open/close hinges. All those vowels and consonants blurring in and out of range during my mother's pregnancy with me, followed by my infancy and early childhood. The way I yearned to fit into the puzzle formed by my traumatized and distracted father, my traumatized and moody mother.

We didn't have words for any of *that* yet. Not for the genetic residue of nightmares and silences, the inheritance of grief and loss. I believed it was my birthright to carry their stories, a form of personal as well as familial evolution. And I listened to my parents much more carefully than they could manage to listen to me.

. . .

What attunement very early in life means for our individual blossoming is part of what is known as "theory of mind": the developmental phase of being able to understand and interpret one's own mental states as well as those of another, especially with regard to the recognition of differences between ourselves and others. It's believed that there are neurological links between this cognitive capacity and messages of interconnection, spoken and unspoken.

Therefore, the absence (or inadequate quantity and quality) of early parent–child attunement interactions can result in damaging effects. At an extreme level, this impact applies not only to the emotional and psychological well-being of children but also to their physical health.

"Relationship rupture is a severe bodily strain," writes Dr. Thomas Lewis, in his book *A General Theory of Love*. "Prolonged separation affects more than feelings. A number of somatic parameters go haywire in despair. Because separation deranges the body, losing relationships can cause physical illness."[16] It's worth noting that this correlation between attachment and well-being applies to mammals as a group, not humans only. "Attachment bonds are a reflection of the limbic architecture mammals share. Short separations provoke an acute response known as *protest*, while prolonged separations yield the physiologic state of *despair*."[17]

. . .

As we age, our most typical loss of hearing affects the upper range of frequencies, those most likely to be emitted by infants in distress. It's as if we aren't expected to hear those cries any longer, once we are no longer at the age of parenting. I try yet again to imagine what it must have been like for both of us, my mother and me, out on the street in front of our home on Van Rensselaer Drive, when I was trying to claim and reclaim her notice.

Is it a good thing that we humans aren't capable of tuning in to the sounds of *all* creatures calling out for their missing nurturers? "Even young rats evidence protest: when their mother is absent they emit nonstop ultrasonic cries, a plaintive chorus inaudible to our dull ape ears."[18] Conversely, in a somewhat disturbing discovery, research recently published in *Proceedings of the Royal Society B: Biological Sciences* indicates that Nile crocodiles "are even better at identifying the emotional cues hidden in the wails of babies than we are—perhaps because they've evolved to home in on helpless prey." According to this new study, the crocs are "uniquely sensitive to the wails of distressed primate babies . . . and the more anxious the cry, the more interested the crocs become." The researchers acknowledge it isn't entirely clear whether the crocodiles were "acting out of parental concern, rather than blood lust." But while the human volunteers mistook some of the distress cries, the crocs "were more likely to respond to recordings with acoustic features known to correlate to highly upset infants such as disharmony, noise bursts, and uneven tones reminiscent of radio static."[19]

. . .

When Grandma Judy came to visit by Greyhound bus from her apartment in Newark, she brought along a vinyl handbag that contained a boiled chicken wrapped in tinfoil; it seemed she would eat nothing else. I could tell from the scalding tones of my mother and her mother arguing in Russian that they were revisiting their favorite wounds and unforgettable losses. Despite my not comprehending a single syllable, I could *feel* their discord, their trouble. Even my father stayed out of it. That agitated dialogue belonged to the two of them, as ever, locked in a tangle of longing and disappointment.

Although I believe my father's anguish must have been embedded in visits to us from *his* father, I don't recall overtly expressed bitterness. As a child I grasped only a vague portrait of a grandfather who, as a father, had left his first wife and three young sons alone in Nazi Germany after the divorce. He remarried and went silent for over a decade, persuaded they'd all been killed in the war. Years later, after living in Israel with his second wife, and yet later, now in a borough of New York City with his third, he and the now-adult sons had been reunited as if restored. After my grandfather died, my father admitted to me that he never forgave his father for abandoning his wife and children, especially in such a dangerous time. Were the explanations hidden inside that forbidden language, the one belonging to the murderers?

To us children he had been simply Grandpa David, source of yet another heavily accented voice entering our house in upstate New York, a tall stranger who shared my family name. He smoked cigars and drank *slivovitz* (plum brandy), lifted me up onto his wide shoulders so I could touch the ceiling. I think now we were all pretending that we belonged together, as if America were home for us all. I knew no other, but they did, or used to. Not only faded, but renounced. Denied.

. . .

Sometimes I think memory is a form of listening. Searching for the lost sound of my mother's voice is an effort to retrieve a file from its hiding place—my mother, a hidden child, my mother who was able to communicate in at least seven different tongues. All those melodies bloomed from her mouth like an audible bouquet. More than twenty years past her death, I want to recover her saying, "My Lizzie," sounding to my ears like "My Leezie," the long vowels echoing the shape of her Polish or Russian, those languages I refused to learn.

I can still recall the sound of her raging in Swedish at us, at Dad: "*Dra åt helvete!*" I guessed at her meaning from hints of association: *Go to hell!* Alternatingly came her shrieking in Polish, "*Choléra yasne!*" Years after Mom died, when I detected familiarity in my friend's new wife's Polish accent, I asked her for a translation. "Bright plague," answered Basia, evidently puzzled by my inquiry.

For the first time, I could make out the connection between her word *choléra* and an English word I also knew: cholera. My mother was wishing a *plague* on us, her children. When I explained the reason for my question, Basia laughed, eyes wide with a kind of shock or maybe admiration, that a mother's rage could be hurled with such intention at her own babies. *Oh yes, it could*, I thought, having chosen definitively by then *not* to become a mother myself. How many times had Mom said *choléra yasne* under her breath, said it aloud, screamed it? Probably her own mother had shown her how, when she was a child herself, the same curse delivered by my Grandma Judy in her own moments of fatigue and fury.

. . .

One summer when I was five, my older sister and I got into an argument over a pebble I found while we were sitting along the curb in front of our house. I was excited about this discovery of what I was certain was a diamond; my sister wanted me to show it to her, and when I handed it over, she wouldn't give it back. Shouting was no use. I ran to the house for help from my mother, who was in the kitchen making dinner and who had (unhelpfully) locked the door, which was (even more unhelpfully) made of glass and not screen, through which I might have explained the injustice of my situation.

I banged and banged to get her to see and hear me. Maybe she was about to shriek *choléra yasne* at me for interrupting her—but I was using so much force that the glass door shattered, and a significant shard embedded itself on the inside of my right wrist. What I vaguely remember next is gushing blood and a panicky ride to the nearest hospital. Ruth, my mother's best friend, who was also my godmother and our nearby neighbor, drove us in her Chevrolet while my mother held me on her lap, my wrist wrapped with a kitchen towel that was turning red fast.

Here is the part I recall most vividly: in the emergency room, as he prepared to stitch the wound, the doctor asked me if I wanted to sit up or lie down. "Sit up," I said, because I wanted to watch this procedure of needle and thread digging into my very own skin. But when he said, "Lie down," I was speechless with confusion and a rising sense of injustice. If he wanted me to lie down, why give me a choice? Why ask me what I want if he wasn't going to respect what he heard in my answer?

Listening is so complicated. It doesn't always make sense.

. . .

The acronym ACEs (adverse childhood experiences) refers to an established ten-item measure that can measure the correlation between childhood trauma (e.g., being raised by a parent with substance abuse issues, having a parent who is imprisoned) and adult health outcomes. The Centers for Disease Control and Prevention found that ACEs so profoundly affect children's developing brains and central nervous systems that they are linked to chronic disease, mental illness, and substance abuse in adolescence and adulthood.

A more recent modification to this system of measurement, called PACEs—positive and adverse childhood experiences—is also being employed in order to consider sources of resilience. Examples of positive experiences that mitigate adversity would include having a family member who offered protection during a crisis, feeling the support of friends, and being able to talk openly and feel heard.

. . .

Recent research on the development of theory of mind in children who are on the autism spectrum has shown that bilingual or multilingual environments can actually aid in the acquisition of language. Although it was previously believed that challenges to such children's communication skills could be mitigated by using only one language in the home, the opposite turns out to be true.

Children with autism typically have difficulties with responding to a question that requires putting themselves in the place of the questioner; the challenge involves concentrating on someone else's point of view—their beliefs, emotions, intentions, and desires. In a bilingual environment, "the child must continually ask him or herself about the specific identity, behavior, and knowledge of others with whom he/she is interacting: 'Does the person I am speaking to speak Greek or Albanian? In what language should I talk to him or her?'"[20]

Our earliest listening practices are intimately tied to our developing an identity, a self, while at the same time helping us form our understanding of others. That is to say, one of the most essential stages of childhood development occurs as we learn that we are simultaneously separate yet interconnected.

. . .

There were nights on Van Rensselaer Drive when my sister Monica and I stayed home in our shared bedroom while our parents were two houses up the street visiting the Englers, an Israeli couple whose son was my best friend. Although I have a specific memory of an intercom propped on the windowsill, this wouldn't have served to connect us with neighbors. Maybe it was meant to reassure us as a visual cue, a symbol of being heard. We were around the ages of six (my sister) and four (me). I imagine my parents wanted us to feel some invisible tether, within reach in case of emergency, to be notified by way of yelling for help.

Correcting my distorted memory now, I sense the intercom doing its job of transmitting and receiving inside our own house. Yet the voices of my parents and their friends also floated up the stairs and through the bedroom door Monica and I made sure to keep slightly ajar, the light leaking in as well as the blended chords of adulthood. Awake and asleep, my ears caught hold of repeated stories and odd jokes, teasing laughter and rustling dresses, ice-filled glasses and porcelain plates tinkling.

These couples weren't alcohol drinkers, any of them, so parties were never about getting drunk or even tipsy. This was about shared history, that old-world residue they each carried, separately and together. *Accents.*

. . .

We traveled to Israel in the summer of 1967, when Monica was nine and I was seven; our one-year-old brother, Raphael, was left with a nanny (astonishing decision!). Joseph (formerly Wolfgang), the brother who survived Buchenwald along with my father, had settled in Bnei Brak, an ultra-Orthodox community located alongside Tel Aviv. My father's mother, Rachel, had spent the last few years of her life there too, dying just one month before my brother's birth. In fact, Raphael had been named in her memory, an approximate arrangement of letters much closer to the original than my own naming, Elizabeth, supposedly honoring my mother's deceased father, Zalman. (In a distant future, a spiraling repetition of sounds would bring my mother's name—Frieda—all the way forward to bless my brother's first daughter, Frida, when she arrives two months past her grandmother's death.)

A big reason for the Israel trip was because my parents wanted to visit Rachel's grave. But when they decided to bring only Monica to the cemetery, leaving me behind with my Uncle Joseph and Aunt Sarah, I wept all day. My uncle and aunt didn't speak English and I didn't speak Hebrew; they couldn't explain I wasn't being discarded. (Hadn't my father and his brother been placed in an orphanage when their overwhelmed mother couldn't manage to care for all three of her children? Didn't I know this?) I can only guess what my little brother must have felt, on the other side of everything, the four of us having vanished without him. It seems evident to me now that we were all primed for separation and loss, the generational agonies repeated and repeated.

. . .

You held your hands over your ears, my mother said. *You refused to listen.*

Psychotherapists would probably call my parents wounded narcissists. For multiple reasons—including but not limited to the shattering of their families in the Holocaust, and the severe disruptions of their war-torn childhoods—both my father and mother suffered in their own childhoods from being unseen and unheard. In simple and also convoluted ways, my parents needed to have all three of their children reflect themselves back to themselves—mirroring their own rare blend of beauty, intelligence, talent.

None of this was conscious, of course. They would have insisted that they were trying to improve upon their own lack of good parenting. They *were* doing the best they could—holding on to us extra tightly, criticizing us with instructions for how to live properly, arguing over the details of those same instructions when their preferences diverged. Listening back now, I can't determine whose voice was more terrifying to my ears: the booming of my enraged father or the shrieking of my fed-up mother. They yelled at us and they yelled at each other, and sometimes all I could think to do was hide for a while in the dark crawl space under the stairs of the house. I waited for some approximate quiet suggesting it was safe to emerge from my cave.

Their past leaked through the wallpaper, carved its sharp edges around and inside the present. Suppressed sorrows and active resentments were amplified by the stern-faced photo albums. I try (and fail) to understand what it felt like for my parents to hear their *own* voices, teaching us in borrowed garments of language, muttering curses from the other side of memory.

. . .

When I think about the accidents, migrations, and coincidences of vocabularies and electricity and history that allowed me to come into existence, I recognize my part in a global as well as molecular longing to belong, an inchoate yet also profoundly specific need to connect with others. Researchers across an ever-broadening variety of fields (biology, cosmology, etymology, neurology) are drawing our collective consciousness to this fundamental developmental requirement for living *with* other living matter. Our emotional, cognitive, and spiritual health as individuals—and communities—are necessarily interdependent.

LIVING IN THE SOUNDSCAPE

"We have no earlids," R. Murray Schafer points out. He is credited with popularizing the term "soundscape" to refer to the sonic environment we inhabit (as contrasted, for instance, with a visual landscape or vista). "We are condemned to listen, but this does not mean our ears are always open."[1] Intriguingly, Schafer distinguishes between what he has called "hi-fi" and "lo-fi" soundscapes.

"The country is generally more hi-fi than the city; night more than day; ancient times more than modern. In the hi-fi soundscape, sounds overlap less frequently; there is perspective—foreground and background." Whereas, in the lo-fi soundscapes, "individual acoustic signals are obscured in an over-dense population of sounds. The pellucid sound—a footstep in the snow, a church bell across the valley or an animal scurrying in the brush—is masked by broad-band noise. Perspective is lost."[2]

. . .

Our specific human range for hearing has evolved to help us focus on sounds that are supposedly essential to our survival, while frequencies too high and too low are filtered out. Every species is evolving with their own ancestral imprinting, of course, yet the sound niches we humanly occupy can—and should—be widened to extend ourselves well beyond anthropocentrism.

In his book *An Immense World*, Ed Yong emphatically encourages such curiosity regarding a wide spectrum of species and *their* perceptions of the world: "Nothing can sense everything, and nothing needs to. That is why *Umwelten* exist at all. It is also why the act of contemplating the *Umwelt* of another creature is so deeply human and so utterly profound. Our senses filter in what we need. We must choose to learn about the rest."

On a call-in radio program during which he was discussing his new book, Yong was asked by more than one listener to comment on the apparent "conversations" we humans experience with our pets, especially dogs. Although he gently declined the request to analyze what might be going on inside the dogs' heads, he did recommend considering the relationship as one of mutual listening. I think of my own years of living with Lulu, a spaniel mix, who convinced me that she possessed a substantial vocabulary both in the form of speaking and also in responding to my speech. Like some other dogs, she repeatedly demonstrated that she could locate by name the individuals among her menagerie of stuffed animals; it was also obvious she knew the difference between our going for a *WALK* or a ride in the *CAR*. But what felt even more significant was my decoding of *her* language. As she aged, I recognized a certain whimper that meant *I can't get down these stairs on my own in the dark*, communications that were audibly different from the ones that meant *Can you help me jump up onto the bed?*

Within our shared *Umwelt*, we had very specific expressions for joyful greetings and sharp warnings, for announcements of love and fear. During our decade together, Lulu and I were continuously listening to each other, learning from each other. The house is now terribly quiet without her.

. . .

"I was always listening," said avant-garde composer and sound artist Pauline Oliveros about her childhood. "Instead of having vivid visual images, I had vivid sound images."[3] Oliveros is credited with coining the term "deep listening." She defined it as "a way of listening in every possible way to everything possible to hear no matter what you are doing. Such intense listening includes the sounds of daily life, of nature, of one's own thoughts as well as musical sounds."[4]

. . .

Andreas Weber, in his book *The Biology of Wonder*, suggests, "We are all insensitive to most of the world's vibrations and energies."[5] Weber is one of many making a case for healthy cognitive, emotional, and even physiological well-being by listening to what Weber calls "this 'other,' another incarnate subject that consists of living matter."[6] If we are to continue sharing the planet more harmoniously, what's required of us is dedicated effort—and engaged listening is key.

Daisy Hildyard, in her essay "War on the Air," writes about the meticulous inquiry we must bring to the presence of sound and also to its absence, evoking echoes of Rachel Carson's groundbreaking work, *Silent Spring*, while highlighting our current cascade of environmental crises. "A person might understand her position in the world not by introspection but by looking out and paying attention to the agency of other things and beings, even when this view diminishes or displaces her right to priority," Hildyard suggests. "A story that extends through the human and the other-than-human, across astronomical and submicroscopic scales, is a story with a widened sense of world. It could be useful now."[7]

. . .

To state it plainly, when we humans get quieter, we can hear so much more. So can everyone and everything else. One of the most illuminating discoveries made during the early COVID-19 pandemic is that the dramatic decrease in noisy human activity, especially in the first year of sheltering in place, greatly benefited numerous species in the wild (and even in urban spaces). In places as distinct as the African savanna and the Pacific Ocean, observers found that the particularly pitched distress calls from cheetah cubs and young whales could now be heard by their mothers across unusually extended miles of desert air or sea water. The extra silence allowed for increased physical distances between mothers and offspring (while the mothers were hunting and foraging, for example). Some endangered species actually doubled their reproduction or survival rates during that year, apparently in response to our human absence.[8]

What cannot be denied is the implication that most of the planet's species—many currently faced with extinction—could be *much* better off without us here at all.

. . .

Acoustic ecologist and self-described "sound tracker" Gordon Hempton, who has been seeking and studying silent spaces on earth for decades, offers yet another variation on the necessity for humanity to decenter ourselves: "When I listen, I have to be quiet. I become very peaceful. And I think what I enjoy most about listening is that I disappear. I. Disappear."

The paradox is that if we have any hope for enduring as a species, in order to avoid or at least defer our own extinction, we must listen more deeply to what Hempton calls "the presence of everything. Silence is the presence of time, undisturbed."[9]

. . .

Gertrude Stein once insisted that paragraphs (unlike sentences) are "emotional"; she found this out by listening to her dog Basket drinking water from a bowl. "And anybody listening to any dog's drinking will see what I mean," she wrote.[10] Although I can't say for sure that I understand what Stein felt and heard while she tuned in to Basket's lapping, I do know that after reading her words, I noticed my particular tenderness whenever I listened to (or even thought about) the rhythmic melody created by Lulu slaking her thirst. Sometimes the pattern was syncopated with the clinking of her dog tag against the edge of the bowl.

. . .

Numerous studies affirm that hearing is not only at least ten times faster than seeing,[11] it's also about ten times more nuanced than vision and the only sense that remains fully operational even while we are asleep.[12] The underlying biological and evolutionary reason for this speed and constancy may seem self-evident: we need an alarm system that never turns off, especially when we are most vulnerable to danger. Despite what feels like an ever-shrinking attention span while awake, most of us bring hypervigilance to sounds our brain interprets as dramatic, urgent, unfamiliar—potentially threatening.

As it happens, "in normal conversation, only a small part of the brain is taxed, leaving excess processing power to be used for listening for lurking predators, filtering out background noise or simply daydreaming."[13] Given such available potential, a relatively recent practice sometimes known as "speed listening" involves adjusting the rate of audiobooks and podcasts to increase your capacity to absorb more material in a shorter time span. While typical rates of comprehension hover around 140 to 180 words per minute, some people dubbed "podfasters" are accelerating their audio consumption by as much as three or four times the so-called normal.

And yet, such voracious appetites for information strike me as the opposite goal from what many acoustic ecologists are promoting when they urge us to slow down and get quieter so we can connect with what matters. This critique from Lauren Murrow in *Wired* magazine states it acutely: "Speed-listeners think they're self-optimizing, but science shows comprehension flies off the rails at 2X—and crashes and burns at 3X. Ever wonder what happened to comfortable silences? Why small talk makes you want to pass out? Why the ring of your phone triggers a Pavlovian punching of the Ignore button? Blame the relentless babble that's blasting your brain to mush."[14]

. . .

The more we observe with our volume turned *down* enough that the natural world's volume is turned *up*, the more we find rewarding guidance for the fragilities of our human condition. A recent TV series called *Super/Natural* (narrated by the irresistibly subtle voice of actor Benedict Cumberbatch) depicts sequences of behaviors that prove instructive for our own species' abilities to survive and thrive within a broader community.

For instance, when nuthatches in a forest send out alarm calls to their own species, their particular frequency also alerts a community-wide cascade of shared news to other vulnerable prey about an incoming goshawk, arriving at predatory speed, forty miles per hour. Burrowing owls have adapted an ability to collectively mimic the threatening sound of a rattlesnake, thus averting consequences from the earthshaking havoc wreaked by buffalo dust-bathing near the opening of their refuge. Acacia trees, in symbiotic relationship with massive ant colonies eating holes in their seedpods, produce a whistling/buzzing sound caused by wind that implies the dangerous presence of swarming bees, repelling elephants who would otherwise consume the trees' leaves.

Clearly these examples aren't only about sound making but also feature highly evolved listening within a networked ecosystem. Thus, the sonic universe can teach and inspire us to broaden the range of our communal listening, to recognize mutual warnings and a fellowship of solutions. As David George Haskell reminds us, in his book *Sounds Wild and Broken*: "At any place on Earth, thousands of parallel sensory worlds coexist, the diverse productions of evolution's creative hand. We cannot hear with the ears of others, but we can listen and wonder."[15] And beyond that state of wonder, many argue, lies a state far more collectively hopeful—not only for humans but for all living things who share this planet with us.

. . .

Born and raised on ranchland in West Marin, California, Susan Hall is an artist whose work I like to describe as a visual representation of her coastal soundscape. In a blog post entitled "The Intimacy of Loneliness," she explores some of the most essential influences on her creative life: "The atmosphere, the ether, is a tuning fork trembling with lives that cannot speak for themselves. The air is as full as a green forest on a dark night. And being alone in wide open spaces is different than being alone in a room." After mentioning the "long horizons" of her childhood, in which she was "steeped in vastness," she adds this: "I have asked questions of fish, clouds, frogs, the sky, and other life forms surrounding me. Then I wait patiently (or not) until impulses or thoughts arise in return. Surprisingly, they often come in snippets or images. Then, I assume a conversation is beginning."[16]

. . .

A BBC Radio podcast called *Slow Radio* is devoted to helping archive some of the disappearing sounds of the natural world. As explained on their website, "It seems that the wonders of our world in sound are as fragile as the planet itself, and just as in danger of extinction as the life contained within it. As our world moves forward ever faster, take a moment to listen—the sounds you take for granted may not be there in the not-too-distant future."[17]

I'm thinking about obsolete (or nearly vanished) sound-linked words from my childhood: words like "rewind," like "dial tone," and the specific reverberations accompanying them. The whirring chatter of a recorded reel-to-reel tape in reverse motion; the sustained note humming before I hung up a heavy telephone; the *kerchunk* of the mimeograph machinery as it tumbled and thumped, awkward and repetitive. Along with audio memory, olfactory and tactile details suddenly rise up: the distinctly peculiar chemical odor of blurry purple ink pushing wet and toxic into my nose, soaking through filmy paper and staining my fingers.

This makes me wonder if people with synesthesia—i.e., entangled stimulation of the senses—ever feel conflicted about how to organize the world. I should note that though estimates vary, the neurological condition of synesthesia is very rare, affecting between 1 percent and 5 percent of the population. In his book *Musicophilia*, Oliver Sacks writes about a man with lost sight for whom "the world of sound is continuously transformed into a flowing world of colors and shapes." In an even more dramatic example, he tells about one of his patients who

> developed a synesthesia so intense as to replace the actual perception of music, thus preventing him from becoming a musician, as he had intended. *I had no sooner made a sound on the A string, or D or G or C, than I no longer heard it. Saw music too much to be able to speak its language. Tones, chords, melodies, rhythms, each was immediately transformed into pictures, curves, lines, shapes, landscapes, and most of all colors . . . I saw music too much to be able to speak its language.*[18]

John Williams, a prolific composer who has won numerous Academy Awards, describes the Abbey Road Studio in which he conducted the London Symphony Orchestra recording the *Star Wars* soundtrack by saying it "has a nice bloom, a nice face."[19]

. . .

"Mismatch negativity" refers to the brain function of listening that monitors our environment for unexpected changes in sound, an early warning system carried over from way back in human evolution. Tragically, the fact that we are hardwired to recognize the strange as dangerous can be manipulated and even weaponized. To cite one deadly example, "parsley"—in Spanish the word is *perejil*—is sometimes used as shorthand referencing "The Parsley Massacre" of 1937 in the Dominican Republic, the test of your "race" having been your ability to pronounce a rolling *r* correctly, the Spanish way. If you were a French-speaking Haitian, you had an accent in Spanish that was either acceptable or wrong. A sound that could save you or kill you.

I can't help connecting this with circumcised boys/men in Europe under the Nazi regime—the clue to Jewishness that could be hidden unless you were literally exposed, a piece of unmistakable visible evidence of your so-called race. Under the genocidal dictator Rafael Trujillo, it was your own tongue that would give you away, the way your words entered the ears of the murderers.

. . .

Dr. Viorica Marian speaks about the "juggling metaphor" sometimes used to describe activity in the brains of people who are bilingual or multilingual. "With language, you have to pay attention to what matters and ignore what doesn't matter," she explains. "To facilitate and prioritize your use of one language, you must inhibit information in the rest. And there is quite a bit of evidence showing that this experience of juggling two or more languages has some advantage for the executive function."[20] Decades ago, I was riding in an elevator with my father in New York City, and a couple standing close to us were speaking to each other in Swedish. Dad and I exchanged a very subtle glance—we both knew that he could eavesdrop on their conversation because of course they assumed that no Americans within range could decipher their language.

Overhearing. Eavesdropping. Tools of my trade, my life.

. . .

"*Har du penge?*" It's a Swedish phrase that occasionally floats up from my childhood audio storage, one of those private questions asked by my mother of my father or the other way around, as she was leaving the house to go shopping. (*Do you have money?*) When I was about twenty, my mother suffered what we only much later called a nervous breakdown, and she said something similar to me on the phone. I picture myself sitting on the scratchy carpeting in my apartment in Palo Alto, holding the receiver that connected me across the miles to her empty house in Vancouver, British Columbia. My polyglot mother was trying *not* to speak English because "someone was tapping the line," she explained, *spying* on her.

My parents had moved across the continent from Schenectady to Vancouver just a year earlier, along with my teenaged brother, because my father had been offered a new job with a Canadian corporation. Almost as abruptly as they had gone west, they were reversing direction to settle in Toronto, closer to what my father had decided was a more suitable spot back east. While my father and my brother drove across Canada to follow their lifetime's collected belongings boxed up in a moving van, my mother remained alone in the home that had been sold. I'm not sure why, but she had been left behind with a cot and a suitcase, to fly and meet up with them in Toronto a few days later.

On the phone, my mother whispered to me that there was "a van parked on her street with a license plate that began DFP," which she was certain stood for *Detective Following People.* Who knew she was so close to the breaking point? Who knew how to read the signs and sounds of her mind losing its way?

She said something to me in Swedish, then in Hebrew, and I, on the long-distance line, didn't know how to reassure her. I understood more than I could speak, in both of these languages. I didn't know what to answer back, which of my meager words she would understand.

"Find the root," my mother used to insist when I struggled with homework, no matter the vocabulary in question, English or Spanish or Hebrew. That discovered meaning always seemed so easy to her, so obvious. Like a mystic, she could see three letters illuminated at the heart of a word, and in listening she could decode their sound with her eyes closed. I guess to her they were like a trail of breadcrumbs in the forest lit up by the moon or the clarity of a murmur reaching across an echo chamber.

I couldn't always see or hear them, though. The roots were elusive, hiding, a mystery. I stumbled around in the silent dark, straining yet half-lost.

. . .

Are most of us simply exhausted from our efforts of simultaneously tuning in and tuning out? What does it mean to fix *all* of our awareness on listening to anyone (or anything)? "Research on listening indicates that we spend about 80% of our waking hours communicating: writing 9%, reading 16%, speaking 30% and 45 to 50 percent of our day engaged in listening, to people, music, TV, radio, etc. About 75 percent of that time we are forgetful, preoccupied, or not paying attention. One of the factors influencing this statistic is that the average attention span for an adult in the United States is 22 seconds."[21]

. . .

One night not long ago, watching a film called *Phantom Thread* on my laptop at my desk in Berkeley, I was surprised to hear the calling of a pair of owls back and forth: three perfect notes, first a sequence of lower ones and then an echoing response of slightly higher ones. I paused the film to make sure I wasn't imagining things, and there they were again, communicating from somewhere in the canyon stretching below my house. Even with my double-glazed windows closed, I detected their elegantly matching sounds while watching the film—which contained a score and plenty of dialogue. I concentrated my active listening directly in front of me, yet my peripheral listening was actively at work too.

The next morning while I was tapping on that same laptop (same desk, same windows facing the California live oak trees) I found myself unable to ignore a steady pounding rhythm from somewhere in my neighborhood. It wasn't exactly a mechanical banging or jackhammering or chopping, but some relentless concussion. At times it seemed as though my body was being hit, over and over. Later, that noise gave way to the whining of a leaf blower. I wanted so much to tune it all out, delete it, and yet I had to surrender to it being there, around me and in some sense upon me, invasive and unavoidable.

Two nights later, I was awakened by the roar of my furnace that had kicked on by some automatic setting I don't remember ever programming into the thermostat. A blast of heated air was filling the house at 3:00 a.m. with warmth I didn't need (in fact it was too much, I had to throw off my covers). I felt as if a voice kept saying *Wake up, Wake up, Wake up,* pushing at me like some uninvited ghost.

. . .

Quite a few writers I know prefer working while their favorite music is playing, or while seated at a café table surrounded by conversation. One of the main reasons I don't choose either of those settings for my own work is that I find it extremely problematic to be pulled in more than one auditory direction at a time. Without ambient silence, it's as though a competition arises between my outer world and the *sotto voce* recitations coming from inside—by which I mean a place that feels both far away and very nearby. In the presence of almost any type of noise, my listening brain can't help but involve itself in processing and interpreting external sounds. And so, my creating brain is essentially hijacked—at least a significant portion of it is. The inner whispers very often lose the fight.

Experts claim that our visual attention affects our listening, so you would think that as long as I keep my eyes on my own page (or screen), I'd be able to disregard the rest. Bernie Krause, in his book *Voices of the Wild,* explains that "our acoustic impressions are often influenced by the other senses. When we're standing on a beach listening to waves, for example, if our gaze is aimed at the breaker offshore, we tend to hear the distant crashes we're looking at. When we're standing near the water's edge looking at the leading wave of the water as it rolls up to our feet, we tend to hear the crackle of bubbles in the surf as the thin membranes give way and release the gases inside."[22]

If that's why closing our eyes helps with concentration of the listening mind, we are still left with the challenge of zeroing *inside* when our eyes are open. Where to look when searching for the voices of the imagination, those almost-real spirits who are trying to speak?

. . .

Evolution notwithstanding, the old lizard brain continues guarding our safety, wired to remain vigilant for the sharply pitched cry of an infant, for the acute sound of any member of our tribe who might need rescuing. And yet, since our brains are organs of learning, we are also designed to be changed by our environments and our experiences. That means we can teach ourselves (and reteach ourselves) how to stay calm in the presence of an alarming noise; we can learn not to panic. What is often referred to as emotional regulation—such as overriding an instinctual reaction—takes effort and practice.

Sometimes I think about the studio apartment where I lived for a few months in San Francisco after completing graduate school in Southern California; perched above not just one but two freeways, I pretended to myself that the roaring of traffic below my new urban window echoed the soothing presence of the ocean, which rumbled a block from my previous apartment.

That is the surf, I told my worried brain, again and again. *Translate this into ocean waves. Into the sound of peace.*

. . .

"I learned to use my senses in different ways," says landscape painter Susan Hall. "Not rating, or calculating or planning, but taking in what was offered to me. These were feelings and waves of interchanges where my senses reached out and melted into the non-human. This is the place where poets stretch their minds and hearts, where the universe offers itself with open palms and empathy reigns."[23]

WHISPERS AND HEALING

LONG BEFORE WRITTEN LANGUAGE, ACROSS MILLENNIA and geography, our human ancestors traded dreams and stories not only as a form of passing on traditional wisdom but also as a means of helping to alleviate suffering. "More than 3,500 years ago, references to 'healing through words' appeared in ancient Egyptian and Greek writings. The word 'counseling' found its way into Geoffrey Chaucer's *The Wife of Bath's Tale* in 1386."[1] Historian Dr. Katherine Harvey notes that people in the Middle Ages confessed their most intimate secrets to priests with a belief they were enabling beneficial treatment for physical, mental, and spiritual well-being. "Medieval hospitals required their patients to confess on arrival, and at regular intervals thereafter."[2]

. . .

My first encounter with psychotherapy began during the summer when I was twenty-two. Barely one year past graduating from Stanford, I was biding my time at a marketing job in downtown Palo Alto while preparing to leave for the start of an MFA program in Southern California. My boyfriend M., with whom I lived, was working on his PhD in psychology, and sometimes I felt I was soaking up his studies by osmosis. (In fact, I edited many of his seminar papers, so it wasn't my skin so much as my eyes and ears that were doing the absorbing.) One of M.'s closest friends and fellow students frequently talked about the creativity and mysticism of Carl Jung; on his recommendation, I began reading Jung's book entitled *Memories, Dreams, Reflections*. And I decided that I wanted to try meeting with a Jungian.

Because of my imminent relocation, my weekly sessions could last for only three months. In agreeing to see me for such a brief period, the therapist said she understood that this was a transition through which I needed some extra support. I have a singularly vivid memory of a dream I told her about—something I now know is considered "an initial dream." I was alone ascending an elevator in a very tall and very narrow building, when the entire structure began to tilt and collapse. Although everything around me was crushed into rubble, I remained unharmed by the catastrophe. As the dust settled, I calmly walked away.

In addition to the dream's sensory and emotional details, I can recall the mood between us when I shared it. The therapist suggested that while the metaphor of the crumbling was accurate, the dream was also reassuring me about my own resilience. Despite whatever I might be fearing about the upheavals of my life, I was going to be all right.

Listen to what your psyche knows. I can almost hear her saying it, elegantly positioning herself not as the source of wisdom but as a kind of echo chamber. Not unlike the dance instructor I studied with more than two decades later, who would often whisper to us through a microphone held close to his lips: *Listen to your teacher. I am not your teacher.*

. . .

Still influential among the most traditional of psychoanalysts, Austrian neurologist Sigmund Freud believed that the analyst's duty was to listen with what he famously termed "evenly hovering attention" (sometimes translated from the German original—*gleichschwebende Aufmerksamkeit*—as "evenly suspended attention"). He was convinced that the effectiveness of the psychiatrist depended upon his or her ability to dispassionately attend to the patient's free associations and intimate disclosures, without judgment but with "calm, quiet attentiveness."[3]

According to Freud, "the rule of giving equal notice to everything is the necessary counterpart to the demand made on the patient that he should communicate everything that occurs to him without criticism or selection."[4] The purpose of such neutrality is for the analyst to avoid anything that might be experienced by the analysand, even unconsciously, as a kind of censorship.

In the famous case history of pseudonymous Anna O., published in 1895, we find the first mention of the "talking cure," a phrase that the poetic Anna herself came up with. "Freud himself once described Anna O. as the true founder of the psychoanalytic approach to mental health treatment."[5]

Starting at the age of twenty-one, she (Bertha Pappenheim was her actual name) was treated for a complex of symptoms by the eminent nineteenth-century physician and neurophysiologist Josef Breuer, who was Freud's mentor at the University of Vienna. In this initial stage of her treatment, Anna/Bertha invented fairy tales and told them to her doctor; sharing them appeared to bring her some measure of relief. The subsequent stages of treatment involved daily sessions of morning hypnosis followed each evening by Anna's recounting of suppressed events and occurrences from the year before—which had presumably served as cause for her symptom onset.

"As she did so, the relevant symptom itself would disappear," explains contemporary lecturer and physician John Launer.[6] He continues, "Most believe that talking works because it provides people with a means of creating a coherent narrative from disconnected symptoms, events, memories and thoughts in the context of a relationship with someone compassionate and attentive."

It's worth noting that Anna's Dr. Breuer is credited with discovering "the action of the vagus nerve on respiration, as well as the function of the semicircular canals." That is to say, Breuer's expertise focused not only on the longest nerve in the body—and the place where vast interest continues to grow regarding the body's storing and processing of trauma—he also studied the vestibular physiology of the inner ear.

. . .

Perhaps psychiatric and psychotherapeutic treatment could be more aptly termed "the listening cure." Whereas classical Freudians might refer to "emotional pollution" in the psychoanalytic encounter (e.g., reactive feelings in the analyst, especially problematic if unacknowledged by him/her as a form of countertransference), Theodor Reik insisted that the psychiatrist's role relied on bringing close observation to the usefulness of such subjective reactions. Awareness of the emotional texture and temperature of what the patient says (and does not say), combined with self-awareness, is key to what has become known as the collaborative approach to healing that Reik promoted, in which both participants are transformed by the process.

Picture the patient reclining on that iconic couch with the doctor seated nearby. Freud deliberately positioned the pair in an arrangement that prevents direct visual contact, with the intention of enabling free association to flow without interference from the listener, even in the form of a raised eyebrow, the hint of a smile or frown. Contrast this with Reik's approach, which has led to numerous variations and continually evolving psychotherapeutic practices, facilitating a shared experience in which both therapist and client explore not only the layers of language being expressed aloud but also their inner emotional landscape for the sake of empathic bonding.

. . .

Most therapists see their role as helpfully interactive rather than strictly interpretive or prescriptive. If listening weren't so critical to the process, we would be divulging secrets to ourselves, alone in front of a mirror, or muttering in the dark corner of a closet. What most of us seem to *need* is another being to take interest in what we are revealing, what we may be in the act of discovering by speaking aloud into the ears of a person we trust and yet may not even know very well. Talking with someone who is fully engaged can actually point us insightfully inward in a new way—not unlike immunotherapy using small doses of specifically targeted substances to boost the healing systems within our own bodies. Third-ear listening can be a rediscovered remedy for our anxiety, our grief, our hopelessness, our terror.

. . .

Through more than a century of psychoanalysis, rooms have filled with multidimensional silences and hovering attention, ears poised like tuning forks of the mind and the heart. As that Freudian couch has mostly given way to an arrangement of comfortable armchairs—patient and therapist both seated upright and face-to-face—Reik's phrase "third-ear listening" has mostly fallen out of common usage. These conscious listening practices are now shared far beyond the confines of a psychotherapy office, with varieties of dialogue based primarily on compassion.

For instance, a movement known as Nonviolent Communication, or NVC, was developed in the 1960s by Dr. Marshall Rosenberg, an American clinical psychologist influenced by humanistic psychology founder Carl Rogers and by Mahatma Gandhi. The essential principles of NVC concentrate not on physical violence but on the voiced or even implied oppressions caused by labeling, criticizing, and judging of oneself and/or others. NVC's widely adapted work has expanded from its initial role in conflict resolution to speaking and listening practices involved with personal growth, education, and more widespread applications in activism for social change. The subtitle of Rosenberg's 1999 book, *Nonviolent Communication*, later changed from *A Language of Compassion* to *A Language of Life*.

. . .

Consider the phrase "pregnant pause." Third-ear listening can be a fertile place, whether we are communicating wordlessly with our eyes or absorbing whispers from across a room or a continent. Perhaps the cultivated and often sacred silences that float inside a therapist's office are not so different from those expectant spaces within our own bodies—filled with potential and possibility.

"Your two ears can take in only so much," writes James E. Miller in *The Art of Listening in a Healing Way*. "They can be attuned only to certain wavelengths. After that, your third ear may be the one that hears best."[7]

. . .

Alongside the most useful generalizations about human cognitive development, there are many varied cultures and subcultures of listening. Here's one example. A Northern Cheyenne writer named Cinnamon Kills First mentions to the moderator of a panel we are on together that she won't just "jump in" to the conversation. Although she is a fierce advocate for a broader and deeper acknowledgment in the so-called mainstream regarding her tribal ancestors' history and culture, she explains that she might need an overtly expressed invitation to speak and therefore feel heard. I wonder how much inner language she must be holding in that place of listening and waiting. Not only waiting until the moment is right and until she has a meaningful offering, but also sensing when the space has been audibly—deliberately—opened up by her listener.

In contrast to this is a practice that noted sociolinguist Deborah Tannen calls "cooperative overlapping," where the person who seems to be interrupting is actually trying to keep the speaker going, encouraging by way of supportive echo. This reminds me of the well-rehearsed practices of a long-married couple I know, where the wife gently but insistently inserts the word "Anyway . . ." to nudge her husband's rambling stories forward. There's a connection as well to the frequent chorus of *Amens* among the congregation of a Black Baptist church, where the speaker is continuously reassured the audience is *with* them, cheering them on, a form of verbal (and often loudly vocal) nodding and agreeing. In yet another cultural context, Japanese listeners use "back channels—interjected responses to a speaker . . . During a speaker's grammatical pauses, which happen multiple times per sentence, a listener will say 'hai'; they might also do so upon hearing something impactful."[8]

For someone unused to this style, or someone sensitive to being cut off and even muted from the start, this intentional cooperation might feel instead like an intrusion, a disconcerting zone in which to struggle to keep track of oneself. At its most extreme, some overlapped speakers might feel so *un*-listened-to that such intermittent responses are experienced as attacks, a conversational door being slammed in refusal and renunciation. Instead of a *Yes*, this could feel like a form of silencing. An absolute *No*.

Among the extensive and ongoing side effects of the COVID-19 pandemic on our personal and collective lives, consider the mute button and the raising of virtual hands in Zoom rooms—not to mention the way current technology quite literally prevents two speakers from "holding the floor" simultaneously. At least for now, conversational overlap appears to be significantly reduced, and there might even be more room for people who tend to wait their turn to speak. "For people who don't particularly like the idea of fighting to be heard, the Zoom era—which creates more orderly queues for commenting during conversation—has been a boon. When the conversation is virtual . . . 'you hear from people that aren't always the best at getting their voices in there.'"[9]

. . .

"Escúchame con tus ojos," a boy says to his papá. *Listen to me with your eyes.* Because he knows what it looks and feels like when he is being fully heard, when his father is giving him directed presence. It's a desire shared by most if not all of us: a longing to receive undivided attention. "Divided" being shorthand for anything less than total contact, an authentic and reliable fusion. *Yes, you matter to me more than anything else, right now.*

Eyes can give it away, as in, they can reveal that the listener is at least partly elsewhere, maybe pretending to listen while glancing sideways, as if wishing you were transparent or not there at all, an obstacle or an inconvenience, a stand-in for whatever else might be nearby or on its way. Voluntary or involuntary, this scattering of regard can feel as distancing as one step out of the room. *Come back,* you might say, with your voice or with your own eyes, pulling and pleading. *I'm here, I'm right in front of you, what could possibly be more compelling?*

. . .

In Tepoztlán, Mexico, in January 2023, on one of my afternoon wanderings around town, I found myself drawn toward a couple of horses I noticed behind a chain-link fence near a grove of enormous laurel trees. There was a pinto and a bay. They appeared to want to greet me too, especially when I offered up some handfuls of fresh grass picked from the side of the dirt road. I guessed that they smelled me before they saw me. And then they acted as if they already trusted me.

It was either coincidence or fate when, at a dinner party just a few days later, I was invited to spend a couple of hours with Tali Gomez Rubio Garibay, the woman who runs Equus Tepoz. She's an equine therapist, and those horses belong to her.

The morning of our appointment, Tali connected with me as the horses did—as though she already knew me, as though we were already friends. Petite, lean, and wiry, clad in dust-covered boots, a T-shirt, and riding pants, Tali welcomed me into her rustic shed/office, where she explained a few general ideas about horse evolution. As she gestured to drawings and charts on the walls, I found her excitement infectious. Soon I was following her uphill to the barn where the horses were eating.

Once there, Tali enacted what could have qualified as a brief but applause-worthy one-woman show. She exclaimed over the electromagnetic fields that we create, and the way that horses generate and sense such fields—"except theirs are eight times as large as those of humans!" Stretching her arms and turning in a circle, Tali said that "when people are inside our house, our energy field, we feel them and also their feelings." She half danced and half strode while counting up to eight, covering so much distance she practically disappeared in the bushes.

"We are *inside* the horse's *house* right now," she announced, pointing at the horses standing about thirty feet away in their stalls.

Tali chose the mare named Pinta to accompany us, and we walked down the dusty trail together, stopping outside the open office doorway. "*Dame tu pierna*," Tali said gently but firmly—not a command but a polite request for Pinta to give her a foot, which she readily did. Tali wanted to show me that the shape of the horse's hoof resembles an upside-down heart.

"The thing is that horses have *several* hearts," she said. "This is how they sense the earth, and also, they stamp their feet to move the circulation up and down their legs." She spoke about how sensitive their soles are, how much they listen through the vibrations of the earth, and why—when they are forced to wear iron shoes—they lose much of that ability.

When Tali described their doubled pairs of nostrils, she abruptly stuck her fingers up the horse's nose to demonstrate; Pinta didn't even flinch. Tali explained about their teeth, their facial expressions, their eyes, and, at last, about their ears. About how as the rider she was always "listening, listening, listening."

Tali demonstrated with her hands posing as ears, from the perspective of a rider looking down at the horse's head from above and behind. She said she was watching the movement of the horse's ears in several different positions, indicating anger, interest, fear. "They have very poor vision, but very strong smell, *very* acute hearing." Then, with a hand on each of Pinta's ears, Tali moved them one at a time to show me how they independently bend in multiple directions, revealing distinct attitudes and moods.

"When the horse kicks from behind it's because she can't see what's back there," Tali said. "But she will let you know ahead of time when she is about to kick. She gives you warnings."

In addition to no horseshoes, there are no saddles at Equus Tepoz. There are bridles but no bits. The horses are "at liberty" often, Tali told me, allowed to wander up the road and into the nearby mountains. A day before my visit, one horse was found to be bleeding because he had eaten some cactus.

Tormenta and Katana were two more horses who came down the hill—after they smelled that Pinta was eating carrots and hay in order to keep her standing beside us while Tali taught me some horse language. When we got distracted, Tormenta stuck her entire torso into the office in search of the bag of carrots. "You found them didn't you," Tali scolded the horse with admiration, shooing Tormenta back out again.

. . .

Maria Montessori, renowned pioneer of an educational approach that centralizes a naturally occurring curiosity, independence, and choice in student learning, wrote: "Children are not only sensitive to silence, but also to a voice which calls them . . . out of that silence. It's as if silence is an action which summons qualities from within the student . . . [and] brings up the knowledge which we had not fully realized, that we possess within ourselves an interior life."[10]

Tali's mother, Laura, had been highly trained as a Montessori teacher, and yet as a young child, Tali simply wasn't much of a reader or a writer. Even Montessori-style learning didn't quite speak to her. But when she was six years old, Tali visited an uncle on the Pacific Coast of Mexico who was training dolphins and orcas. In the water with her uncle and one of the orcas he was working with, she realized that this was where she felt not only comfortable, but entirely at home. Suddenly, immersed in this nonhuman environment, she realized she had found her place in the world, her mission, her path. From there she extended herself toward birds, rabbits, cats, dogs, and finally toward horses.

"This is not just my work, it's my language," Tali said, smiling.

. . .

"Horse whispering" refers to a form of gentle and cooperative training that emphasizes rewards over punishment, reassurance and reinforcement over control and "breaking." In the Western world, the modern term appears to be traceable to an Irishman named Daniel "Horse-Whisperer" Sullivan, whose methods of rehabilitating traumatized horses became widely known in nineteenth-century England. Subsequent popularity in the United States and elsewhere broadened after the phrase appeared in books and films, although there are plenty of horse trainers who overtly reject this terminology.[11] Increasingly common nowadays is the concept of "horse listening," with a focus on attunement through subtle communication flowing back and forth between human and horse.

While some trainers rely on what they call a nonverbal, silent conversation between rider and animal, Arizona-based Navajo horse trainer Jay Begaye describes a vocal lineage directly inherited from his sheepherding mother. As a toddler, he sat behind her on horseback while she sang.

"I remember putting my arms around her waist with my ear pressed against her back," Begaye says. "I could hear her songs, but more than that, I could feel them." He explains that his songs are "the same ones used two hundred years ago," and that "a trainer who sings from the heart earns the horse's trust and respect."[12]

In training clinics hosted by Begaye, together with New Mexico-based trainer Boyd Brodie, the preservation of Navajo culture and imagery is promoted through reflecting on the horse's "place in emergence stories (a horse embryo was among the gifts from the Holy People) and the symbolism of the horses' bodies."

. . .

A practice known as therapeutic horseback riding is being developed and studied with regard to particular benefits for autistic children. Researchers are finding that the child's emotional regulation is measurably improved through subtle interactions between human and horse that balance calmness with attentiveness. These effects are similar to those resulting from mindfulness practice—which is often used therapeutically to help autistic children. Such benefits from what might be considered a form of third-ear listening also reduce the need for medication and hospitalization.

"This nonverbal communication between the horse and the rider may include the fact that horses constantly mirror and respond to the rider's body language."[13] And like children on the autism spectrum, horses often show a distinct preference for repetitive routines and habits. Studies also suggest that "a horse's rhythmic stride can have a calming effect on the brain."[14]

As if to prove the point in her own way, Tali, in Mexico, shared with me a few ideas about the evolutionary connections between horses and humans, how we walk in similar ways, how the rider can regulate with the horse, synchronizing our movements and our heartbeats and our breathing. "Children who have epilepsy or autism can have their nervous systems calmed almost instantly," Tali said, reflecting on her personal research at Equus Tepoz. "They can relax."

. . .

Popularized as the Horse Boy Method, a treatment for neurodivergent and disabled children is now being used in more than thirty countries. The approach was originally discovered almost accidentally by a British-born horse trainer named Rupert Isaacson, when his two-year-old son, Rowan, was diagnosed with autism. Children on the spectrum typically overproduce cortisol, a stress hormone sending the nervous system into fight-flight-freeze activity; the gentle rocking of being on horseback appears to stimulate production of the so-called happiness hormone, oxytocin, which not only calms the amygdala but makes room for the neuroplastic brain to focus on learning. Two decades on, Rowan, who had once been assessed as permanently nonverbal and dependent, enrolled in college, owns his own car, and lives independently. It's estimated that the Horse Boy Method "helps around 300,000 people per week, including German army and US Air Force Veterans."[15]

. . .

In the first house of my childhood, the one on Van Rensselaer Drive in Schenectady—the place where we planted maple trees and neglected the backyard garden—I often crossed the dark hallway in the middle of the night, tiptoeing from the bedroom I shared with my sister to the one my parents slept in. I didn't dare look downstairs into the shadowed living room, where our sofa and armchairs crouched like hungry beasts. Although I'd been startled awake by terrors I might have swallowed from my parents' haunted childhoods, I sought comfort in their reassuring arms. I tried to describe to them what I'd seen and heard, the vivid disasters that threatened to erase me.

Shhh, my father whispered. *Go back to sleep. Just a dream.*

. . .

Anne Karpf writes that "interpreting other people's inflections and modifying our own is one of our most important interactive tasks.

We voice-read to confirm what our other senses have told us, and sometimes use the voice to express feelings and moods that, if put into words, might leave a trail of embarrassment or shame."[16] Ideally, what we are continually practicing are subtle yet essential exchanges that occur through minute and almost-invisible, inaudible dialogue.

And, as seems so often the case, research is catching up to the intuitive knowledge so many cultures, ancient to modern, seem already to possess. Studies in the growing field of collective neuroscience confirm that "the experience of 'being on the same wavelength' as another person is real, and it is visible in the activity of the brain."[17] This measurable alignment of brain patterns is occurring between teachers and students in classrooms, between partners in romantic relationships, and among close friends. As for that collective euphoria you feel when you're attending a live concert? When an audience is listening to a musical performance, neurons are firing together in synchrony, and "the greater the degree of synchrony, [one] study found, the more the audience enjoys the performance."[18]

. . .

"The not-said is as important as the said," writes Anna Deavere Smith. "Yet not saying is not the same as lying, it is not the same as covering. In authentic speech, it is what is felt that is transmitted."[19]

Across several decades of performing, teaching, and publishing books, Smith has embodied the voices and stories of her multicultural, multigenerational subjects. During interviews with thousands of people, she relies on an especially keen ability to listen with a finely tuned presence for the breaking up of rhythms, the shift, the emotional places that she came to call "the time that somebody starts singing." For her most recent work, *Notes from the Field*, which focuses on what has been called the school-to-prison pipeline, Smith interviewed over 250 people, then very selectively chose nineteen to portray in depth.

"It could be that I know how to read the water, when someone is talking to me," she says, referring to a woman she interviewed who was describing her practice of trout fishing. "I'm looking for good stories; I know people have better stories than I have. And I'm there to engage with it if it's gonna come forward."[20]

I'm quietly struck by a story Anna tells me about her grandmother's advice to her mother in the presence of a wailing baby (one of Anna's much younger siblings): "Let 'em cry, it's good for their lungs. Let 'em cry."

She explains further: "I am not afraid of another person's pain. I'm not afraid of being in the presence of it, at all. It's not that it belongs to me, but I'm perfectly comfortable in the position to hear that, the dignity of that. I don't want to turn away from it."

We talk about the ways that this willingness to sit with someone's suffering is what can be considered a form of healing, but Anna declines to call herself a healer. Instead, she focuses on the concept of "radical hospitality," inspired by Dutch theologian Henri Nouwen. Nouwen wrote about listening as "spiritual hospitality," in which listening not only to friends but also to strangers is a way of inviting and welcoming someone inside your heart, inside a shared silence.

"How do we welcome the perfect stranger?" Smith echoes. "When I feel I've accomplished something, it's when I know I'm a perfect stranger. And this person I'm listening to is a perfect stranger. The state of the perfect strangeness, and then the connection, is part of my quest as a human being."

Smith believes that "art can inspire action," and in the sharing of *Notes from the Field*, she hopes to help audiences reimagine a world of learning that serves all of us in a more just, respectful, and inclusive way. "But that is a type of reimagining that needs to include all kinds of voices," she urges, "especially those that have been historically discounted. It is a reimagining that requires courage, empathy, and action. And it has to start with listening."[21]

. . .

As we live with the persistent physical and mental health effects of the COVID-19 pandemic, much remains to be learned about the role that psychotherapy can play—including the pivot to remote offerings necessitated by quarantines and lockdowns. In 2020, around 41.4 million adults in the United States received treatment or counseling for their mental health, a 50 percent increase from 2002.[22] Even in the early days of the pandemic, experts were predicting a dramatic uptick in the widespread needs for support, some going so far as to anticipate a worldwide mental health crisis with an indefinitely long aftermath. One article in the *American Journal of Psychotherapy* speculated this way:

> Even as the first wave of infection passes, stressors associated with COVID-19, including self-quarantine, social distancing, job loss, and threat of illness, will persist. These factors can be expected to have a significant impact on the human psyche and contribute to a secondary mental health epidemic. Extrapolating from studies conducted in the aftermath of other disasters, such as Hurricane

> Katrina, Hurricane Sandy, and the terrorist attacks of September 11, we can expect as much as 10% of the population to meet criteria for major depressive disorder following a crisis and perhaps even more to meet criteria for posttraumatic stress disorder.[23]

By March 2022, the World Health Organization announced that in the first year of the pandemic, "the global prevalence of anxiety and depression increased by a massive 25%."[24] The report also indicated a disproportionately large impact on women and young people. We are still in the early stages of discovering how remote psychotherapy can enhance as well as challenge the practice of listening when connections between patients and those treating them must stretch across telephone lines and digital platforms.

Every day seems to bring a new or updated study about the benefits and limitations of remote therapeutic treatments—with no clear consensus yet, given the range of mental health issues and treatment modalities under consideration. Here is part of a summary from one study of 217 therapists in Austria (not coincidentally, Freud's birth country):

> Remote therapy offered the respondents more flexibility in terms of space and time. Nevertheless, the therapists also reported challenges of remote therapy, such as limited sensory perceptions, technical problems and signs of fatigue. They also described differences in terms of the therapeutic interventions used. There was a great deal of ambivalence in the data regarding the intensity of sessions and the establishment and/or maintenance of a psychotherapeutic relationship.[25]

. . .

Ever since that first experiment in my early twenties, for much of my adult life I have spent innumerable hours in one therapy office or another, including most recently a series of sessions via Zoom and FaceTime. As the aforementioned study notes about remote experiences, while I appreciated being able to meet with my therapist without having to leave my home, I also felt the occasional strain of garbled communication or a moment of lost subtlety due to dropped or glitchy signals. More intimate in terms of setting because my own private space was being viewed, I felt both more relaxed and more vulnerable; I also felt more confusingly distant, distracted by the uncertainties of technology and the awkwardness of seeing myself being seen.

Was the virtual contact better than nothing? Yes.

Although not continuously and not with the same therapist, over several decades, I worked my way through the end of a marriage, the death of my mother, many milestones of failures (and even some successes). At one point when I was in my mid-thirties, my father sent me a book entitled *When to Say Goodbye to Your Therapist*, and I fumed at him over the phone for the presumptuousness of the gift. A year later, however, out of sheer curiosity, I cracked it open and learned, within a few pages, that it's believed the completion of therapy is itself a therapeutic process.

Epiphany! I was stunned to consider this possibility, and brought it up with my then therapist, realizing at the same time I had imagined she was the one who would decide when I had "done enough." That revelation allowed me to entertain the possibility of my own self-assessment. It turned out that she wasn't holding some future better version of me in mind, waiting for me to catch up or grow up; it was up to me, and to us together, to proceed along that path until we agreed on a point of arrival. I think now that this final phase of work we shared was, in some ways, the first time I let go of a meaningful relationship in a truly conscious and healthy way.

. . .

"Conversation is the vehicle for change," writes Terry Tempest Williams in her book *When Women Were Birds*. "We test our ideas. We hear our own voices in concert with another. And inside those pauses of listening, we approach new territories of thought."[26] I love that she uses the phrase "in concert" to describe this collaborative creation; it's no coincidence that we often *compose* our ideas and feelings by talking about them not only with therapists but also with partners, friends, even with "hospitable" strangers. We discover new territories by adventuring, out loud or in a whisper, beyond the edges of what we think we know.

THE SPACES BETWEEN

When I was eight years old, attending Hebrew school two afternoons a week, Mr. Benjamin Friend taught us the importance of praying in a low mutter so that we could hear ourselves communicating with God. We were reminded to say *Hashem* (the Name) instead of *Adonai* (God) unless we were genuinely reciting a prayer rather than practicing or even referring indirectly to Him.

For as long as I could remember, I wanted to understand what it meant to be "religious" on the inside, and how I might become more filled with light. Following my teacher's demonstrations, I meaningfully chanted the repetition of *Kadosh, Kadosh, Kadosh* (Holy, Holy, Holy) while stretching up onto my toes with each syllable. Unlike bowing at the waist as we had already been taught in order to show our humility—this striving gesture would demonstrate our devout wish to aim closer to God, where the angels hover. Nobody had to know that I was comparing it with my ballet lessons with Miss Danzig, when the *relevé* extended not only my feet but my entire body, when I imagined becoming my longed-for taller self.

These instructions were provided in a gentle yet scholarly voice by Mr. Friend, a sweet, white-haired man with, I believed, the best of intentions for us all. Except one unforgettable day, my heart shattered when he began to teach the boys the daily morning prayer that translated as "Thank you, God, for not making me a woman," while teaching the girls to say, "Thank you, God, for making me as I am."

The teacher explained that this was not to diminish the girls but to remind the boys that it is a privilege as well as an obligation to praise *Hashem* every single day, a *mitzvah*, while the girls are spared this expectation because we are primarily responsible for the care of children and the household. *Mitzvah*, he wanted us to understand, is both a commandment and a blessing. There are not only ten of these *mitzvot*, Mr. Friend insisted. There are 613 of them altogether.

Words of protest remained trapped in my throat, a sudden pain I could barely choke down. I couldn't fathom why the spiritual shape of the life I had been granted was being restricted without my permission, why I was meant to believe that conversations with God were being determined by my gender. Something I didn't select but was born into—yet something that was, according to Mr. Friend, unconditionally constrained. I'll never forget the way the boys in my class snickered and puffed out their chests, recognizing that they were being chosen to approach God from a higher and more special position. Here was their evidence, in holy print and in the spoken holy words. The wound cut deeper as I saw the other girls slumping in their seats, as if in sullen acceptance of our secondary status.

I wasn't merely hurt, I was angry. The injustice was obvious even to my eight-year-old brain: Girls are supposed to feel *respected* by our presumably sacred task of motherhood, our separate-but-equal assignment? That night at dinner, when I tried to explain my feelings to my father, he insisted that these prescriptions were not to be taken as anything "negative." I was baffled by my sister's agreement with him; in fact, she seemed to *like* going to Hebrew school. My mother, leaning against the sink, shook her head with what I thought was a hint of pity. Did she see the gleam of tears in my eyes? Why didn't she say something?

Instead, my mother shrugged in surrender to *what is*, helpless to rewrite the rules. Oh, there were so *many* rules, and who was she to modify them on our behalf? My father made things worse by asking who I thought I was to question these wise men who devoted their entire lives, centuries of them, to the study of Judaism. He reminded me that I was just a young girl with so much more to learn about the ways of the world, complicated laws written by men adhering to the word of God, the Name, the Holy One Blessed Be He.

All those male pronouns were not lost on me either, not to mention how in Hebrew, every verb has to be matched with a gender, which meant that every action by God was clearly an action depicted in the form of the Masculine, in the form of He, Him, His. And according to the teachings of my Orthodox tradition, for a *minyan*—the quorum of ten participants required in order for a prayer service to be considered official—only adult men are counted. If there are nine men and one hundred women, there is still no *minyan*. The outrages kept multiplying.

So many things were being made clear to me by way of language, audible scripts and manuals for the body I inhabited in a shared-yet-divided world. In synagogue, females were required to sit on the side, or in the back; these separate sections were delineated by solid barriers or opaque screens to keep us "hidden" from the gaze of men, our presence a distraction from their prayers. We were not allowed to go up to the front to read from the Torah, because Holy Scrolls and the inscribed Holy Words are not meant to be touched by our unclean hands. We would be reminded again and again that our bodies are made to bleed, and therefore we must purify ourselves each month in the ritual bath, the *mikvah.*

Not that anyone I knew had ever gone to a *mikvah.* Our family's Orthodox customs were somewhat compromised—as in, diminished—by our remote location in upstate New York, and only a few married women in the synagogue appeared to be wearing wigs. (This had to do with yet another confounding notion: that a woman's hair was either meant only for her husband's eyes or to be shaved entirely so as to allow her scalp to be cleansed in the *mikvah* too.)

My mother certainly didn't wear a wig; in fact, she defied so many rules that she preferred to go shopping while my sister and I walked to synagogue with our father on Saturday mornings. Even in winter, we trudged through the snowdrifts, covering the mile or so between our house and the recently built brick building that happened to be located across the street from the Catholic school. Although I was allowed to wear pants beneath my dress to add warmth for the wintry walk, I was required to remove them before I entered the sanctuary. Where I had to sit on the side.

Awash in a stream of lessons, suffering what I received from my elders, nevertheless, I was itching to find out where else I might listen for what was true. My father accused my mother of setting bad examples, as their voices pitched against each other with more and more vitriol. In addition to my hiding place in the crawl space under the stairs, sometimes I stood for an hour or more in the green-tiled stall shower downstairs. I rinsed myself in the consoling cascade of water, searching for my own way of feeling purified.

. . .

"You must leave behind your expectations and commence your journey," recommends musician and philosopher David Rothenberg. "Or find the journey in the places you have already gone, in the sounds that define the arc of your whole life until now. Only then will you gain the skills necessary in order to hear the future."[1]

. . .

At nine years old, I lay belly down on the dense beige wool of our living room rug, chin propped on my fists. I was listening to *The New World Symphony*, Symphony no. 9 in E Minor. My father had told me that astronaut Neil Armstrong took along a tape recording of this symphony during the Apollo 11 mission, the first moon landing, in July of this same year, 1969. With half a billion other people on earth, I got to stay up late watching the blurry images on our black-and-white TV, hearing a voice carry across space: "That's one small step for man, one giant leap for mankind."

As I started to read the back of the record sleeve, the wordless music pouring into me began to contain something readable, a map. *Adagio. Largo. Scherzo. Finale.* Four movements. Magic portals framed what I was listening to, building a design language in my mind. When I turned my head to extend my ears one at a time toward the stereo speakers (those fabric-covered rectangles outlined in grainy wood that later I would learn are concealing things called "woofers" and "tweeters"), I was climbing inside a tapestry of stories. *New World.*

How many times did I play that record? I can't say. But I remember how the album cover's biography of Antonín Dvořák, the Czech composer, told me about the influences he encountered during a visit to America—a quilt of folk song and Quaker hymn, spirituals of the South, wide-open prairies in the Midwestern landscape. Although these notes referred to several places I had never laid eyes on, it struck me that the United States wasn't just the country of my birth and nationality but also a richly composed territory that could be imagined and loved by an outsider.

So, I listened more deeply for clues to my own origin story, the American newness that offered itself to my parents as Holocaust survivors. Refugees from the *Old* World, their former homes and extended families and even their given identities erased and destroyed, they had no choice except to board ships and cross oceans. I thought about how they planted those backyard fruit trees, the ones with promises for baking. I thought about my father's gaze, aiming at the future, and about my mother's fluent tongue, stretching to accommodate yet more vocabularies, syllables she would need to sing a chorus of lullabies. Both of my parents closed chapters of dust, and then they wrote their own symphony.

. . .

In addition to their ever-growing collection of books in English and a few other languages, my parents collected record albums, mostly classical. Together with *The New World Symphony*, our boxy stereo played *Peter and the Wolf*, op. 67, joining birdsong with flute, duck with oboe, cat with clarinet, wolf with French horns. Here was a grandfather's voice as a bassoon, and young Peter with his own melody of violins and cellos. John Gielgud's sonorous voice helped explain the fairy tale—and I was stunned to notice that in the very last phrase, the duck could be heard "inside" the wolf.

What would *my* signature instrument sound like? Was I made of brass or wood, portrayed by strings or percussion?

Like Terry Tempest Williams, recounting her own childhood in *When Women Were Birds*, I replaced the needle on Sergei Prokofiev's composition again and again, the narrator helping me with what Williams calls my "first tutorial on voice. Each of us has one. Each voice is distinct and has something to say. Each voice deserves to be heard. But it requires the act of listening."[2]

. . .

While something termed the "Mozart Effect" has been mostly debunked, many parents around the world persist in playing classical music for their infants and youngsters, in case it might have some extrinsically valuable influence on children's cognitive development.[3] It may be more accurate to say in broader terms that *emotional* development can be impacted by our earliest exposure to music of all kinds. My heart rises up every time I hear even a few bars of *The New World Symphony* or *Peter and the Wolf*.

That sense of being returned to some original experience of listening is something James Wood writes about in his *New Yorker* review of a book by English poet Patrick Mackie entitled *Mozart in Motion*. For adolescent Wood, the music he repeatedly played was the last piano concerto by Mozart (no. 27, K. 595), especially particular passages from the Larghetto, which he believed were perfectly suited to his temperament. "I was thirteen," says Wood, "fundamentally cheerful but convinced I was fundamentally melancholy." After describing the movement in elegant detail, he questions his own memories. "That passage still provokes my tears, but they are not of grief so much as of gratitude, tears while smiling. Perhaps, then, it is the perfection of this beauty that moves me, with no specific emotion expressed by the notes themselves?"[4]

. . .

In sixth grade, although I wanted to join the chorus at school, my parents emphatically said *No*. In a rare moment of solidarity, they agreed that this was not an "appropriate" activity for their daughter. When I demanded an explanation, my father reminded me that I could sing with my sister and brother on Friday nights at the dinner table and at services on Shabbat mornings.

One of the only things I enjoyed about our Friday nights was the singing of *Grace After Meals*, especially the sections in which my sister and I harmonized. This marginal glimmer of happiness was almost entirely absent when I was in synagogue. Nearly all of the prayer melodies sounded mournful and monotonous to my ear, except for a few rare moments when I could hear my voice blending and yet holding its own alongside the voices of the men. One prayer in particular, about the Tree of Life, ascends into a harmonic sequence so sincere and beautiful it brought me to tears every time I sang it. I had a good voice. I wanted to use it more. And all my friends were joining the chorus.

My parents exchanged a look; maybe a word or two passed between them in Swedish.

"You can't go to after-school chorus practice because you have Hebrew school," my father said.

"Also," my mother said, "we don't want you to sing Christmas carols."

Ironically, one of my mother's stories about being pulled out of high school by her parents involved her wearing a friend's necklace featuring a tiny gold cross. Having survived the Vilna ghetto's liquidation and years of hiding before being liberated, my mother and her parents were awaiting visas in Sweden while my mother attended a girls' school in Stockholm. When my grandparents noticed this shiny Christian symbol being worn by their teenaged daughter—quite possibly as a gift she'd received from a friend, or out of an innocent wish to belong—they quickly reenrolled her in a Jewish school for orphans and refugees. Which was how she ended up meeting my father.

. . .

When I was twelve, my parents sent me to an Orthodox summer camp for one month, the same place my sister Monica had attended the previous year, a place she adored and in which she flourished. After returning home with even more fervor for the precise rules of keeping kosher, she scolded our mother for being too lax when she mistakenly used the same sponge for washing pots used for milk dishes and those for cooking meat. Insistent upon every single detail for observing the Shabbat, my sister was newly committed to the interpretations of "rest" that included prohibitions against driving, cooking, cutting, tearing, carrying, writing, handling money, and using the telephone. In order to avoid igniting or extinguishing any flame (including electricity), my sister began affixing tape over the light switch in the bathroom so it would remain in the ON position from sundown on Friday to sunset on Saturday; she also taped the switch in the refrigerator to the OFF position so that it wouldn't be activated every time we opened the door. Monica prayed three times a day at home now, holding her small prayer book, facing toward the east. My father hoped these transformations would happen to me too.

Within my first days at summer camp, I noticed the way the studious *yeshiva* kids—who comprised the vast majority of my fellow campers—recited prayers at such speed that it was clear my part-time Hebrew school classes were no match for their full-time Jewish educations. It was clear that they all knew the prayers by "heart." Still, to my ears the muttered words sounded almost heartless, a stream of syllables without meaning or soul; it was as though everyone besides me was engaged in a race to finish first. The problem wasn't that I *couldn't* keep up with the pace, it was that I wasn't sure I wanted to. I watched the girls and boys *davening* faster and faster, all day and especially on Shabbat, their adolescent bodies bobbing as if to express devotion but without seeming to have authentic feeling for the prayers' actual sacredness.

To make matters worse, I felt stranded in the intense humidity. All of the girls were made to cover up; the details had been enumerated in our camp packing lists. "No sleeveless shirts, no shorts or skirts above the knee." Why? Because we posed a distraction to the boys at their prayers, a source of temptation. As for the blessing of water? Girls and boys were forbidden to enter the lake at the same time; separate hours were designated for these subtribes. Too much skin, too much dangerous desire, too many revelations of arms and thighs. Our bodies were verging on womanhood now. We were told this concealment was a form of power, but I recognized the messages of powerlessness all over again.

In mid-July we were instructed to fast for twenty-four hours in honor of *Tisha B'Av*, a holy day marking the anniversary of the destruction of the ancient Hebrew temple. Moreover, we were prevented for an entire week from swimming in the lake, prohibited from games or sports as an act of collective mourning, and also a way of submitting our pious bodies to God. Near the end of my month of camp, we headed out for an overnight in the woods, tramping through the mud; *schmutz* it was jokingly called, instead of *chutz*, hike. While my sister had made friends for life at this camp, I didn't find a single girl who shared my alienated bitterness, my sorrow. Secretly, I planned my own immersion in the cool quiet of some other lake, maybe even an ocean, somewhere far away. In awe of the dripping canopy of pines and oaks as we climbed higher into the mountains, it amazed me to realize that the counselors and campers all seemed to agree that nature represented a series of trials and obstacles to be endured if not fully dominated—including the nature of my body. For the time being, I gave up trying to resist out loud. I would learn how to pray with my own words, my own silences, and in my own voice. Maybe I was separate from everyone except the other nonhuman creatures in the forest. Inhaling the scent of wet earth, I privately conversed with the rustling leaves and listened with my heart for the holiness of birdsong.

. . .

"Many bird species are not melodic combatants but lone dreamers enveloping themselves in a veil of song," writes Andreas Weber. "Individuals of species like the garden warbler chatter in a special species-specific low-voice melody only when they are alone and undisturbed. They are talking to themselves, delicate and melodious and totally free of utility."[5]

. . .

Despite the overnight camp, or maybe because it failed to change me, my father and I yelled at each other more than ever. We couldn't seem to talk without raising our voices, pushing sharp-edged words back and forth through the air. But we were especially loud on the weekends. Especially because of Shabbat. Because of rules.

A typical fall Friday at dusk: My father dressed in his suit and tie, having just returned home from work. Standing in the kitchen, I could see into the dining room where the pair of plain white candles were already on the table, waiting to be lit. In front of my father's seat, an empty *kiddush* cup and a plate with fresh *challah*, covered with an embroidered velvet cloth. The bread knife my parents had owned since they lived in Sweden lay just beside it, along with a small black prayer book.

I was trying to explain once again why I hated the obligatory ritual of Saturday mornings, why being in the Orthodox *shul* made me feel simultaneously ignored and judged, unwelcome and put-upon.

"I hate sitting on the side with all the women!" I shouted.

My father had heard my distress before, and although he never seemed to listen, I recited it again anyway. I hated that I was supposed to wear a dress, even in winter, even when we had to trudge through mounds of snowdrifts to get there. I hated that I wasn't allowed to go up to the front to read from the Torah because the rules said I was unclean. I hated that the boys and men were thanking God every single morning for not making them women, while I was supposed to thank God for "making me as I am."

My father shouted back. "Of all the religions of the world, Judaism is the best one for women. The most respectful."

I shook my head, tossing my messy hair around. "What do you know about it!"

We glared at each other. The more we repeated ourselves, the higher our volume, the less we understood. Turning away from me, my father stepped into the dining room to pick up the prayer book. "You have to study it more," he said, pointing at the book like he wanted to force its contents directly into my brain. "Judaism is thousands of years old!"

He told me he would always be older than me, always smarter. I yelled back that he didn't know what it was like to be *me*, would never know. How could he possibly imagine how I felt about being his daughter, about being anyone's daughter. At the stove, my mother was adding more salt to the chicken soup; my sister and brother were elsewhere in the house. The tablecloth's pristine whiteness made me angrier. I longed to grab the square bottle of Manischewitz Concord Grape wine and smash it, staining everything.

"Why is every rule about saying *No*?" I pleaded, my rage mixing with sorrow. Why couldn't my father recognize how hurt I was, how the very rules he most loved were the ones breaking my heart.

I was beginning to see why people talked about their blood boiling. My skin felt like it might burst into flame, and my hands were clenched so tightly I could make out two sets of fingernails digging grooves into my palms. I was mystified by the clear evidence that my sister loved being more and more observant; the differences between the two of us widened year by year. My brother was too young to care about any of these arguments. And how could he become my ally? He was a boy, learning to repeat those prayers just like the rest of them.

"The rules are *mean*," I sputtered. "They're old and obsolete and biased against women. They don't *apply*."

A frown deepened on my father's face, his heavy eyebrows knitting closer together. This expression gave me a flicker of hope, as if he might finally be interested in learning more about my point of view. Maybe he would lower his voice so I could lower mine too. Behind me, I heard my mother humming something, an old-world melody. She interrupted her humming to point a long-handled wooden spoon at me, the same one she had used for hitting at my arms when she was furious for reasons of her own, when she was erupting out of control, cursing me and what must have felt like her miserable life. Except now her eyes were crinkling with amusement. "You really should be a lawyer," she said.

I didn't want to be a lawyer—nor did I want to be a doctor, the alternative future she often recommended. I didn't want to be anything other than who I was, the person I chose to be. My parents took turns reminding me that I only existed because they made me, and they acted as though that gave them the right to keep creating and controlling my life.

It's not fair, I announced, over and over.

Life isn't fair, they said back, every single time.

. . .

"The oldest flutes that have been dug up are forty thousand years old, and human song long preceded that."[6] In a *New Yorker* article by Jaron Lanier, son of a concentration camp survivor and self-described "compulsive explorer of new instruments and the ways they make me feel," he writes that "flute teachers I've known have talked about 'blue' or 'yellow' air flows. I've had long conversations with wind players about how we seem to be painting the flow of air inside our bodies."[7]

. . .

Within a few years after those Dvořák and Prokofiev symphonies, instead of joining the school chorus, I began learning to play the flute. Sweet birdsong was translated again into an instrument, except now it was something silver I could hold in my thirteen-year-old hands and fill with my breath. My teacher, Miss Greene, gave lessons in the Stockade area of downtown Schenectady, a place with a deeply layered history: colonial settlers and a defiant legend about an Iroquois massacre. (It would be decades before I dug beneath the textbook version, when a descendant of the Mohawk tribe told me the true story, explaining who massacred whom.)

Miss Greene said she always recognized my footsteps in her hallway, her senses amplifying mine. "You have a distinct rhythm of your own when you walk," she said. "Do you hear it?" Whenever we played duets, our notes interwoven, I could barely maintain my *embouchure* because the harmony we created was so elegant and sublime. My lips wanted to stretch too wide, into a smile.

Eventually, I was proficient enough to study a solo piece by Claude Debussy called "Syrinx." Two pages of dense music contained a message about danger as well as transfiguration, because the flute held a story inside *its* body. Debussy's hauntingly gorgeous composition both challenged and elated me, a dozen flats and sharps embedded within complex rhythms I wanted to *feel* more than measure. Miss Greene encouraged me to buy a metronome like hers, so that I could count the beats with accuracy.

At my recital (there is a single photo, taken by one of my parents), I was sixteen and wearing a dusty-blue peasant dress I adored. My unruly hair was pulled back into a ballet bun fastened at the nape of my neck, and I was feeling tall on my favorite wooden-heeled platform sandals. Addressing the audience, I was serious as well as transported: before I played the first note, I took a few minutes to explain the passages from Ovid's *Metamorphoses* about the nymph Syrinx, a follower of Artemis, being chased by Pan until she reached the edge of the river. When she pleaded with the river nymphs to be spared from capture, they changed her into reeds. And it was the wind blowing through her arms and her hair, now hollow, that inspired this melody, these strains of mystical escape.

. . .

Although I continued playing the flute on and off for many years, I always limited my repertoire to classical pieces or the work of contemporary musicians I loved. It wasn't until a friend encouraged me to experiment with improvisation that I discovered how very nervous I could feel in the presence of a sound I couldn't predict or, at least partly, control. Convinced I was incapable of inventing music that didn't already exist, I believed such abilities relied on a talent or imagination I lacked.

. . .

Referring to a song that became a kind of psychedelic anthem in the late 1960s, musician Grace Slick said once, on social media, "I wrote White Rabbit on a red upright piano that cost me about $50. It had eight or 10 keys missing, but that was OK because I could hear in my head the notes that weren't there."[8]

. . .

Keith Jarrett, one of my all-time favorite musicians, described how he creates space for his improvisational performances on solo piano. In order to "walk on the stage and play from zero" he has to "*not* play what's in my ears, if there's something in my ears. I have to find a way for my hands to start the concert without me . . . As the music is being played—I am very aware. I have prepared my awareness. I can respond swiftly to the whole broad range of what my ears tell me can happen."[9]

. . .

John Cage's premiere of his composition *4′33″* (four minutes, thirty-three seconds) occurred on August 29, 1952, in Woodstock, New York, not very far from where my parents would soon be settling in Schenectady. An article about Cage's premiere describes it this way: "Unlike compositions designed to make the outside world fall away, here was a music that, when it engaged you, made the present world open up like a lotus blossoming in stop-motion photography."[10]

Terry Tempest Williams writes about that 1952 concert too: "In times of war, survival depends on listening to that suffering. Cage understood how the unexpected action of deep listening can create a space of transformation capable of shattering complacency and despair."[11] Cage himself once said, "There is no such thing as an empty space or an empty time. There is always something to see, something to hear. In fact, try as we may to make a silence, we cannot."[12]

. . .

"I wish there were a way to make 'I don't know' a positive thing, which it isn't in our society," Keith Jarrett has said. "We feel that we need to 'know' certain things, and we substitute that quest for the actual experience of things in all its complexity."[13]

. . .

The balance between questing for knowledge and bowing to not-knowing is especially relevant in the world of improvisational music-making. Renowned inner ear surgeon Dr. Charles Limb has spent more than two decades studying the neurobiology of musicians composing in real time. In a TED/MED talk from 2011, Limb explains that functional MRI (fMRI) research has shown—in the case of individuals who are considered master improvisers—decreased brain activity in regions responsible for self-monitoring and increased activity in regions related to self-expression. For example, one of his case studies focused on brain scans of Gabriela Montero, a Venezuelan-born pianist whose repertoire combines classical and improvised programs. Limb reports that improvisation is "a highly engaged, unique brain state" involving "auditory, frontal/cognitive, motor, parietal, occipital, and limbic areas."[14]

Even while working with large quantities of data in pursuit of greater insight and understanding, Limb acknowledges that we actually know very little about the mechanics of the brain, that we are just barely beginning to make sense of its complexities. A jazz saxophonist himself, Limb explains to me that he is grateful and happy to have managed to combine his passions for music and medicine, after deciding he wasn't "good enough" or "courageous" enough to attempt to make a living as an artist. Then he remarks that he is leading a more musical life as a surgeon than if he had tried to become a professional artist, because even his surgery feels musical to him. When I ask what he means, he says that his focus is on restoring hearing so that his patients can "hear the voices of their loved ones, so they can listen to music."[15]

Gabriela Montero is quoted in Limb's study as describing her own improvisational state as "constantly creating music without trying," and "constantly hearing and composing music 'in her head,' even when speaking to others." She "likens improvisation to a game," and "is largely unconscious of what happens during improvisations . . . Improvising for her is similar to 'turning on a faucet' and that she can do this at any time without prior preparation." This seems a particularly apt metaphor for what is often called a "flow state . . . the heightened optimal experience of deep focus, creativity, and enjoyment that often accompanies creative tasks."[16]

. . .

I confess to remaining unclear how much is voluntary or involuntary about this process of creativity in the brains of those Limb calls "eminent musicians who are creativity outliers even among elite professional musicians." Yet I believe it might be suggested in general terms that our willing surrender to *not-knowing* makes third-ear listening possible, leading each of us not only to our own artistic possibilities but also to revelations about the creations and co-creations and re-creations happening inside and all around us.

. . .

Clarinetist, professor, and composer David Rothenberg has spent decades listening to the music of nature, becoming a collaborator with nightingales, whales, and even cicadas. "In improvisation," says Rothenberg, "you can hear the greatest personality of the performer, more than his notes or his instrument." He is not speaking of humans. "Since the early 2000s, he has taken his clarinet and saxophone to some of the furthest corners of the planet. The result is a new form of music that invites us to question where art ends and science begins."[17] Rothenberg's videos, recordings, and books document many of his remarkable observations and experiences of interacting musically with the natural world.

"To deeply listen to a nightingale is to feel the power of a musician who is not human," Rothenberg writes. "A purveyor of sounds as ancient as they are futuristic."[18] In his book *Nightingales in Berlin*, he describes remaining spacious while playing live clarinet with these birds: "I want to listen as much as the bird does. We don't fight each other for attention—we strive for mutual comprehension."

. . .

Kathleen Dean Moore, in her book *Earth's Wild Music*, reports on robins who first imitate, then "invent and improvise as freely as Ella Fitzgerald."[19] The twentieth-century French composer Olivier Messiaen also "had a lifelong fascination with bird song, notating and including bird melodies in his music. He once wrote, 'I doubt that one can find in any human music, however inspired, melodies and rhythms that have the sovereign freedom of bird song.' He mentions blackbirds and nightingales in his preface to the *Quartet for the End of Time* [composed in 1941], and while he doesn't specifically assign an instrument to a particular bird, the violin does sing like the nightingale and the clarinet the blackbird."[20]

. . .

Lou Fancher, writing about composer, violinist, and music educator Wendy Reid, explains that "hundreds of the world's finest modern composers live right outside our doors and windows. Making their homes in trees, bushes, wheelbarrows, under porches or roof awnings and in rare instances underground, birds and raptors arguably represent nature's most innovative music makers. Every day, even in urban areas, a person highly attuned can hear a virtual aviary concert taking place in the open air, entirely free of charge."[21]

. . .

Within the Western world, recognition of the cooperation between human composition and specific birdsong goes at least as far back as Wolfgang Amadeus Mozart, who—in May 1784—encountered a starling in a Viennese shop singing a distinctively improvised variation of one of his piano concertos. After making the bird his beloved pet, Mozart spent three years (as documented in diary entries and letters) relating to the starling as a companion, influence, and muse. Bird-watcher and naturalist Lyanda Lynn Haupt, in her book *Mozart's Starling*, writes about this history as well as her own relationship with a bird whose capacities extend far beyond mimicking words (*Hi, honey!* and *C'mere!*). Carmen, the starling, constantly imitates the environmental sounds of Haupt's household, specifically timing these announcements not in reaction to but in anticipation of their occurrence. She is deliberately connecting aurally with those in her social context.[22]

. . .

Rothenberg and Haupt and Moore and Reid are not the only contemporary artists and observers turning their curiosity toward the musical compositions of the nonhuman environment. "Today, in a relatively new field known as soil bioacoustics—others prefer terms such as biotremology or soil ecoacoustics—a growing number of biologists are capturing underground noises to open a window into this complex and cryptic world. They've found that something as simple as a metal nail pushed into the dirt can become a sort of upside-down antenna if equipped with the right sensors. And the more researchers listen, the more it becomes apparent how much the ground below us is thrumming with life . . . We humans might be among the last to discover this underground soundtrack."[23]

The researchers involved in studying soil acoustics aren't just doing this for esoteric reasons. Think back to the groundbreaking work done by Rachel Carson, in *Silent Spring*, that led to massive reductions in our use of DDT and other toxic pesticides. "Scientists are also finding that the underground orchestra of animal activity has started to fall silent in large tracts of land, particularly in intensely farmed fields, where 'things go quiet,' says [Zurich-based sound artist and acoustic ecologist Marcus] Maeder. A lessening of noises hints at diminished biodiversity . . . That dovetails with a recent report by the Food and Agriculture Organization finding that a third of the world's land has been at least moderately degraded, often through agriculture. Maybe soil acoustics will help more people realize what we're in danger of losing."[24]

If you wonder whether you are only imagining how much less there is to hear in nature: "There are 30 percent fewer songbirds in the forests now than there were fifty years ago . . . Frogs have declined 38 percent. In Europe, 30 percent of the cricket and grasshopper species are threatened with extinction."[25]

. . .

As a flute student, it was challenging to read music if I hadn't already heard it first, especially if the piece was rhythmically complicated. I was very good, however, at learning quickly by ear, reproducing what I'd heard. Decades have passed since I've played it, but I can remember the first several phrases of "Syrinx." My fingers may not play the notes as fluently, but I can still hear them in my mind's ear. And then I recall the way the final sounds of the piece are meant to fade out, like final exhalations, like dying.

. . .

We humans are faced with choices about how to interpret the music of the planet, whether warnings and/or blessings. Rothenberg, for one, seems able to retain optimism in his own practice. "I always try to avoid hearing the sadness of a lament when listening closely to the interconnectedness of our vanishing world. I don't really want to believe it is vanishing. With all the bad news that confronts us about nature, from rising seas and temperatures to the loss of thousands of species every year, it is important through music, art, and sheer experience to take in all the sound around us, to be better listeners so that we realize just who and what we all are."[26]

. . .

David G. Haskell's sonic journey "When the Earth Started to Sing" includes references to oceanic mammals such as whales and seals whose "larynx returned to the water, and sang."[27] Whales "are more closely related to the camel and cow than any fish in their wake. Their anatomies retain vestiges of the four-legged land animals in their ancestry, the ones that began the return to the sea more than 50 million years ago."[28] Evolutionary biologists clarify that whales migrated back into the ocean to exploit greater opportunities for feeding. "Other changes over time involved improvements in the animals' maneuverability in their new environment. These included hearing systems evolved for underwater communications and navigation and completely new ways of locomotion, first as amphibians and then as fully developed marine animals . . . Whales in their present form began appearing about 30 million years ago."

"Sound's role has always been to connect," Haskell says. "Despite being fugitive and ephemeral, sound is generative. To listen, then, is to be open to the vitality and creativity of life.

"Yes, we are often destroyers, but we are also sonic marvelmakers. Our music can join mind and emotion to other people and species. The spoken word can unify us into collective action. Listening expands the bounds of kinship, joy, and right action."

. . .

Human-generated noise can be a wall that inhibits or even prevents so many different types of interactions, inner and outer, promoting din instead of spaciousness, interfering with the voice of the self as well as the music of the natural world. (Even though we are part of the natural world we also, often, defeat it.) Our arrogance is sometimes matched by our lack of curiosity, but eventually, if we last long enough, we might learn.

It's paradoxical that we coevolved with so many species whose vocalizing is beneficial to our own survival. "We hear most acutely in the range of 2.5 megahertz, which is the peak of birdsong. Human speech is pitched much lower, one kilohertz or below, and so is less central to our hearing. Why is this? Acoustic ecologist Gordon Hempton surmises that our bodies evolved not for cocktail party conversation but rather to harvest sounds from wild creatures. These are the aural signals on which our species' success depended: Birds chatting, unconcerned. Herds gathering. Corvids flocking. Sudden silence. They spoke clearly: *Here is safety. Here is water. Here is food. Here is danger.*"[29]

. . .

"Sound and silence are not opposite, as many people mistakenly believe. They work together," says Wendy Reid.[30] She has been living and collaborating with Lulu, an African gray parrot, for more than seventeen years. In an article about Reid's performances, she explains that "I bring my bird to concerts and people sometimes say they didn't hear her do very much. But I don't train her to have to perform. She is there listening and that is valuable. People I work with understand silence and how to allow it in their improvisation."

. . .

LISTEN and SILENT are anagrams in the English language. A man named John Francis spent seventeen years in complete wordlessness, elucidating later in his book: "Listening is so important, because without listening we can recognize neither silence nor each other." At the conclusion of his extraordinary process, he chose Earth Day to begin speaking, "so that I will remember that now I will be speaking for the environment."[31]

. . .

Although there was a Unitarian Society in my hometown, and historical landmarks noting old settlements established by Quakers and Shakers, it was decades beyond my childhood when I learned about their religious practices involving listening. Whether alone or in a group, whether tuning in to something called by the name of "God" or "the still small voice" or no names at all, these rituals promote listening inward for guidance and for light. Quaker gatherings might consist entirely of silence, when any message to be shared with the group is first deeply reflected upon, a sense of waiting not for permission but for feeling led by some inner voice to speak aloud. Silence is only broken when someone is "moved to speak," not in the form of rules or dogma or doctrine, but in humble offerings of wisdom.

When I was invited to speak at a Quaker college several years ago, the host informed me just prior to the start of my presentation that, in their tradition, any time a group gathers together, the speaker begins with silence. In giving myself a minute or more to breathe deeply along with the soft breathing of my audience, I felt awed by this reminder to stay present, to gaze wordlessly into the sea of upturned faces in front of me, and to hope in all humility for the capacity to offer something useful. Even if that usefulness only happened for a fleeting moment, for one person in the room, even if I would never know the reception or impact of my words.

. . .

In the Buddhist practice of group sharing, it is customary for a speaker to "bow in" and hold the floor until he or she "bows out," followed by three full breaths to allow for space in between, thus ensuring a full allotment of time as well as absorption, time to integrate what has been said before what might next arise. Renowned Zen Buddhist teacher Thich Nhat Hanh speaks about the necessity of heartfelt listening that embraces all beings and even the planet itself: "What we most need to do is to hear within us the sound of the earth crying."[32]

WE ARE STARDUST, WE ARE GOLDEN

An ancient tree may have saved my life. Canadian scientist Dr. Suzanne Simard was inspired to do her groundbreaking research in forest ecology while undergoing treatment for breast cancer. One of her chemotherapy medicines, Taxol, was derived from Pacific yew trees thanks to their production of a chemical defense system for their preservation against disease. My own treatment for breast cancer in 2009—an invasive disease I most likely developed as a result of a genetic mutation known as BRCA that I inherited from my mother—included a variation on this same drug, known as Taxotere.

Although for decades her ideas were considered to be far outside mainstream science and most views and practices in forestry, Simard is now recognized as helping foster global understanding of what she calls the "wood-wide web." Her studies have turned "understanding of forests upside down, from just a collection of individuals competing with each other to this entwined, interactive suite of species that actually collaborated and cooperated. They had this whole way of conversing with each other that was complex."[1]

Simard credits this knowledge as being long held among Indigenous peoples for thousands of years. Forest management involving "selective harvesting, controlled burns, an appreciation of the forest as a diverse community—is far wiser than western profit maximization." She also contrasts Indigenous practice with industrial clear-cutting and monoculture plantations. "First Nation friends 'have whole words that describe how trees communicate and feel,' which capture non-animal perception better than English can."[2]

Research by Simard that had been previously dismissed in the 1990s is now inspiring esteem for the ways that interspecies sharing is essential to ecosystem survival. Her more recent work revolves around what Simard calls hub trees, or Mother Trees: "the biggest, oldest trees in the forest—the ones prized by loggers." Connected via mycorrhizal fungal systems, these mother trees are not only sending energy and nutrients to nearby seedlings but also ensuring the vitality of the interconnected forest overall. For instance, "when a tree is dying, it passes on most of its carbon through its networks to the neighboring trees, even different species."

A professor of forest ecology, Simard began conducting experiments with colleagues and students in British Columbia, discovering "the trees were also receiving messages that increased their defense against the beetle and other disturbance agents in the forest and increased the health of those next generations." Arguing that the removal of supposedly unhealthy older trees prevents such regenerative networking, she explains that what she calls "kin recognition" is vital to the future success of entire forest ecosystems. "I measured and analyzed," Simard explains, "and saw how the forest gives forward, passes forward."[3]

Spreading her message via TED Talks and podcast interviews, Dr. Simard's language is no longer deemed "inappropriately" anthropomorphic. "Let's listen to ourselves and listen to what's known . . . We're all one, together, in this biosphere, and we need to work with our sisters and our brothers, the trees and the plants and the wolves and the bears and the fish. One way to do this is to start viewing the forest in a different way: that, yes, Sister Birch is important, and Brother Fir is just as important as your family."

. . .

The medicinal value of trees doesn't only depend on their extracted botanical chemistry. We humans can benefit from something as basic as ensuring their direct proximity to our bodies. As practiced for centuries in Japanese spiritual traditions, "*shinrin-yoku*, a form of health care and healing in Japanese medicine, literally means 'taking in the forest atmosphere,' or 'forest bathing,'" writes Bernie Krause. "Developed during the 1980s, this work has become a cornerstone of preventive health care."[4]

Nearly ubiquitous these days, in articles stretching from *Vogue* magazine to website promotions from health care giant Kaiser Permanente, you can find recommendations for mindful presence while immersing yourself in nature. The relevant point of forest bathing is not to retreat for days and weeks into the wilderness but to spend ten or twenty minutes each day outdoors, whether in an urban park or a dense wooded area. *Listen to the wind and taste the air*, advise the experts.

Garret Keizer, author of a book entitled *The Unwanted Sound of Everything We Want*, reminds us that America is "the place Oscar Wilde once called 'the loudest country that ever existed.'"[5] Whether you are attracted to or annoyed by trends regarding self-help practices, "it should come as no surprise that noise carries a number of adverse and well-documented physiological consequences: deafness, tinnitus, high blood pressure, heart disease, low birth-weight, even statistically significant reductions in life span."[6] Now scientists are validating the healing power of spending time in nature—once again, knowledge held by many Indigenous cultures for millennia. "One particularly magical finding is that phytoncides, the aromatic organic compounds that trees emit—a.k.a. the stuff that makes a forest smell so very good—boost our immune system. Phytoncides help protect trees from pests and pathogens, and, it so happens, protect us, too: When we breathe them in through the forest air, they increase the number of natural killer (NK) cells in our body."[7]

. . .

It's equally noteworthy that a decade or so after the Japanese government first incorporated forest bathing into its national health program, the Japanese Environmental Protection Agency (in 1997) "launched a program called *One Hundred Soundscapes of Japan: Preserving Our Heritage* . . . They include such audible treasures as the 'Drift Ice in Ohotsku' (No. 1), which during the months of January through April 'creates strange, out of this world sound,' the 'Cranes of Idemizu' (No. 97), 'making a great symphony with their singing and beating of wings,' the 'Sound of Wood-Engraving from the Cobble-Stone-Lined-Town of Inami' (No. 54), the 'Strong Yet Elegant Old Steam Locomotive of Yamaguchi Rail-line' (No. 78), and the 'Cicada Chorus of Yama-Dera [literally "mountain temple"]' (No. 4)."[8]

. . .

"The song of the indri is an unearthly sound," writes David Quammen in his complicated study of extinctions, *The Song of the Dodo*. "It carries through the forest for more than a mile . . . It has been said to be one of the loudest noises made by any living creature. It's a sliding howl, eerie but beautiful, like a cross between the call of a humpback whale and a saxophone riff by Charlie Parker."[9]

What is meant when we call a certain quality of sound "otherworldly" or "unearthly"? As if it is so unfamiliar that it must originate from beyond the edges of "our" planet, our recognizable expanse of the supposedly known. Thanks to the longevity of Homer's epics, we can read about Greek sailors from thousands of years ago, albeit with a slightly updated understanding of what they might have been listening to. "Deep down in the hulls of their ships they thought they heard mermaids chant," Andreas Weber comments, "when it was in fact whales flooding their nocturnal ocean space with their acoustic presence."[10]

. . .

In an interview with Krista Tippett for her popular podcast, *On Being*, zoologist and acoustic biologist Katy Payne talks about her lifelong listening practices. Payne, along with her then husband Roger Payne, were among the first scientists to discover that humpback whales compose ever-changing songs to communicate. In Bermuda in 1968, a Navy engineer named Frank Watlington "had been monitoring an array of hydrophones many miles into the sea," when he realized he was hearing sounds made by whales.[11] Fearing that whalers might use this information to facilitate their hunting, Watlington was keeping his discovery secret, until he chose to share it with the Paynes during their visit.

Katy Payne reports: "Well, the ocean is really huge. When you get out on a little boat, you know it. You're clinging to a cork . . . And out there, rolling around and swimming through and perfectly at home in the waves, are these enormous animals. And by golly, they're singing, of all things. They're doing something that we recognize as singing. And so what that has done for me is to make me feel that what lies ahead to be discovered is absolutely limitless. We are not at the pinnacle of human knowledge. We are just beginning."[12]

It was Payne's persistence that enabled her to decode the massive data gathered over a period of decades. As one colleague described it, "she had the patience and tenacity to pore over 31 years of recordings, making cutouts of the musical lines spewed out by a spectrograph, painstakingly analyzing and overlapping them, and listening over and over again until she could discern the individual voices in a whale quartet and then write out the score."[13]

I wonder if Katy Payne ever listened to the transformative silence of John Cage, or if she had any idea that placing her full scrutiny on whale-songs was going to change the art of music and the art of listening to the world, forever.

. . .

Thirty years ago, while visiting the island of Maui, I spent a morning snorkeling with about a dozen other tourists around the edge of the Molokini crater. The captain of the catamaran had warned us to avoid floating anywhere near the unforgiving current known as "The Tahiti Express," because if you got tugged into its grip, there was no escape until Tahiti. In other words, *Pay attention and stay close.*

After a few ecstatic hours of staring underwater at multitudes of fish, we were back on board when the captain announced there were humpbacks nearby; he dropped a hydrophone into the water, and the air suddenly filled with what he assured us was whale-song. My mouth opened to say *Really?* with such a doubtful attitude about this "live" moment that the captain pulled the cord of the mic up and out of the sea, spoke into it, and then dropped it back into the turquoise water. No further proof needed.

I recalled a record one of my friends had played at a party, back in the 1970s, around the same time I was also learning how to meditate. Someone at the party said Katy and Roger Payne's *Songs of the Humpback Whale* had already sold millions of copies around the globe. Hearing these haunting melodies directly from the sea below our small boat made my skin tingle and my heart flutter. It occurred to me that the captain's hydrophone had been a kind of portal into another world. Maybe it was time to believe my own ears.

Later the same day, I went swimming at a nude beach fondly known as Little Makena. I was in the water wearing nothing but a mask and snorkel, when I spied a languid sea turtle. As I happily dove below the surface to follow the *honu*'s graceful journey, my newly trained ears picked up the recognizable music of the whales. I could hear the wavelengths traveling toward me—across one mile? two? ten?

Somehow, I hadn't until that moment understood that I already possessed this ability. On the catamaran with the hydrophone, I believed that special audio equipment was required—that like a dog's ears able to detect frequencies too high for human reception, this resonant music couldn't be listened to in any way that was "natural" to humans. But there I was, in my unprotected skin, no device except the tiniest of bones inside my body. A labyrinth with fluid and membranes and waves of pressure.

The sea turtle swam farther away, its flippers reminding me of wings. It looked like an aquatic ballerina from *Swan Lake*, or a submerged angel. The haunting music kept coming, and it seemed I was eavesdropping on a conversation among behemoths barely plausible in their majestic size and power, their language meant not for me but for each other. Rather than vulnerability, I felt humility and gratitude toward all I was given to witness, even if it was more than I could comprehend. Coming up for air, diving down again, I listened in a state of awe as the universe expanded, widening out beyond any realm I could name.

. . .

"The soundscape underwater has been called 'acoustic daylight' for the way it reveals things about the world that cannot easily be seen."[14] This is what Callum Roberts explains in his book *The Ocean of Life: The Fate of Man and the Sea*. And Kathleen Dean Moore wonders, "How many generations taught how many generations to sing songs so compelling [to humans!] that they outsold the Monkees?"[15]

When David Rothenberg played clarinet with humpback whales in Hawaii, he said it was "like improvising with a Japanese musician."[16] He further commented, "Not everybody is going to think it's music. Not all the whales are going to like it. Not all the humans are going to like it. It's an attempt to cross the limits of our species using art, using music, using sound." A French-made television documentary features him on a boat out at sea, with hydrophones dropped down into areas where the whales are nearby. "The main thing is to listen to what's going on," Rothenberg says to the camera. "If you want to interact with an animal musician, you want to play something that leaves space for them to be singing and joining in, and not just play your music, without regard to the situation but with an open musical mind, to give a lot of space."

. . .

As Carl Safina writes in his book *Becoming Wild*, "so deeply did the whales' music reach us that a recording of humpback whale singing is among the few sounds included aboard the *Voyager* spacecraft. Humanity's calling card to the galaxy has taken humpback song beyond our solar system. It's a message in a bottle—humankind hoping, perhaps, that an alien life-form of great and cultured intellect will understand."[17]

. . .

Selected for NASA by a committee whose chairman was astronomer Carl Sagan, "the Voyager Golden Record contains 116 images and a variety of sounds. The items for the record, which is carried on both the *Voyager 1* and *Voyager 2* spacecraft . . . are natural sounds (including some made by animals), musical selections from different cultures and eras, spoken greetings in 59 languages, human sounds like footsteps and laughter [Sagan's], and printed messages from President Jimmy Carter and U.N. Secretary-General Kurt Waldheim."[18]

Among the musical selections are *Songs of the Humpback Whale*, Chuck Berry's recording of "Johnny B. Goode," "Dark Was the Night, Cold Was the Ground" by Blind Willie Johnson, Bach's Brandenburg Concertos, and Beethoven's Symphony no. 5.

In one segment, described as "Life Signs," there is a recording (compressed from one hour to one minute) of the heartbeat and brain waves of author and documentary filmmaker Ann Druyan (who eventually married Sagan). "Hooked up to a computer that turned all the data from [her] heart and brain into sound," she describes her approach this way:

> I had a one-hour mental itinerary of the information I wished to convey. I began by thinking about the history of Earth and the life it sustains. To the best of my abilities I tried to think something of the history of ideas and human social organization. I thought about the predicament that our civilization finds itself in and about the violence and poverty that make this planet a hell for so many of its inhabitants. Toward the end I permitted myself a personal statement of what it was like to fall in love.[19]

. . .

In 2006, when cosmologists began actively searching for distant signals in outer space, they were profoundly puzzled by the discovery of a signal six times louder than they had predicted, a sound that came to be known as the "space roar." The instrument they used had been built by NASA to "extend the study of the cosmic microwave background spectrum at lower frequencies, the Absolute Radiometer for Cosmology, Astrophysics, and Diffuse Emission (ARCADE)."[20]

Nearly two decades after the discovery, the source of the "roar" remains unknown . . . and there is even debate among scientists as to whether the source is located inside or outside the Milky Way.

. . .

"The human body consists of about 70 percent water and our bones are great conductors of vibration," writer Sharon Heller summarizes. She has spent years studying the physiological effects of sensory stimuli impacting the human body, especially that which is not only irritating but potentially dangerous for those who suffer from sensory defensiveness disorder, or SDD. "The lower the sound, the more our joints and bones pick up the vibration. As sound is made up of air pressure waves that hit our bodies, we literally 'feel' sound. Once a part of the body vibrates, the blood cells carry this resonance to the whole body very quickly."[21]

Although I am grateful not to be afflicted by SDD, I was forever changed by that day I heard whale-song underwater. There was the me I was before and the me I became afterward. What if the "window of universal sound processing" found in the newborn human brain has something in common with that acoustic window inside whale brains? Before we walked on land, we were once creatures of the sea, after all, and prior to being born we were afloat. I like to imagine we were always listening for the multilingual story of the world.

. . .

In Rachel Howard's *San Francisco Chronicle* review of a program by the San Francisco Ballet, she quotes choreographer William Forsythe, who calls ballet "a way of hearing." This presents the idea of dancers as exceptionally embodied listeners, trained not only with refined muscular grace but also with particularly skillful attunement to the relationships between musical space and physical space. This makes me think of certain sign language interpreters who *dance* music so that a Deaf audience can *see* their movements as a form of hearing. ASL is already a language of the body, but this is rhythm made visual, along with mood and essence.

As Deaf writer Ross Showalter explains, "Signs are made, not just with hands, but with fingers, arms, chests, jawlines, faces . . . And, depending on the context and the emotional climate, a sign could either be delivered languidly or quickly. A sign could be delivered, heavy with tension, the hand vibrating with emotion."[22]

Howard ends her review by referencing Susan Sontag's assertion that "the most urgent function of art is to help us 'recover our senses.'"[23]

. . .

In an interview with artist-photographer Tony Gonzalez along with his favorite longtime model, Shannon, we speak about the ways they listen to each other in their work, a relationship that has lasted across decades. Shannon tells me about the wordless noises Tony makes when he is shooting and looking and waiting and seeing; Tony tells me about the subtle movements Shannon makes before and during and after she hears him. There has already been extensive dialogue back and forth for hours beforehand, informal conversation during which they are planning and preparing the setup, the environment, the narrative, the themes. Then, on the day of the shoot: they are creating together in some realm that is both spoken and unspoken. A shared experience mutually understood through her skin and the light and the lens of his camera. Along with what might be called the third-ear listening that connects them.

. . .

Perhaps it shouldn't be surprising that whales have much in common with elephants. Both species can be described as "giant mammals with long life spans who form matrilineal pods." Both species "range over vast distances, and they can communicate at frequencies below the level of human hearing, with sounds that travel for miles; they are extremely social and can express joy and curiosity."[24] Not coincidentally, bioacoustics researcher and whale-song recorder Katy Payne was one of the first to understand vocal/audio similarities between the two species. That acoustic window in the anatomy of whales is also found in the feet of elephants, as they listen to vibrations transmitted through the earth—not unlike the way sound waves are transmitted through water.

Elephants are the largest of land mammals, whose maximum weight can exceed thirteen thousand pounds (the African species, that is). As described by Lawrence Wright in a *New Yorker* article, "their huge brains are capable of complex thinking—including imitation, memory, coöperative problem-solving—and such emotions as altruism, compassion, grief, and empathy."[25]

In Katy Payne's book, *Silent Thunder: In the Presence of Elephants,* she offers a gorgeous and sometimes heartbreaking depiction of her decades-long study of elephant communication.

Starting with her own physical recognition that vibrations she felt in the air could be related to the messages nearby elephants were sending and receiving, Payne began recording what her ears couldn't yet detect. After speeding up the recordings, she was able to hear and measure patterns and unmistakable correlations to individual as well as group behaviors. The elephants create and receive low-frequency rumbles that can keep them connected to one another across miles of land. They are literally staying "in touch" through sound.

It's also not a coincidence that Payne is a practicing Quaker, and thus steeped in deep listening. Contemporary Quaker poet Philip Gross has said: "I was brought up bilingual, in English and in silence."[26] Robin Mohr, in a YouTube video about theological language in the Quaker community, says: "I think that listening in tongues is our spiritual tradition of listening beyond the words for the message of the Holy Spirit in the words that fallible, ordinary human beings have to use to communicate."[27]

In her interview for *On Being* with Krista Tippett, Payne says: "Just being silent is a most wonderful way to open up to what is really there. I see my responsibility, if I have one, as being to listen." She reports:

> Elephants do something marvelous that I wish we would do more of the time. This is something you do find in Quaker meetings, actually, and in Buddhist meetings as well. The whole herd, and that may be fifty animals, will suddenly be still, completely still. And it's not just a stillness of voice, it's a stillness of body. So, you'll be watching the moving herd. They'll be all over the place; they'll be facing all directions, doing different things. Suddenly everything freezes as if a movie was

> turned into a still photograph, and the freeze may last a whole minute, which is a long time. They're listening. When they freeze, they tighten and lift and spread their ears. This, among other things, tells us that they're concerned with what's going on over the horizon.[28]

. . .

Lawrence Wright's article "The Elephant in the Courtroom" discusses a longtime effort to achieve legal "personhood" for an elephant named Happy, who has lived in captivity at the Bronx Zoo for more than forty years. Extensive studies—on elephant self-awareness, group behaviors, and communication suggesting complex thinking and feeling—are being invoked by lawyers in order to help make the case that Happy (and by extension, elephants generally, as well as—here's the huge challenge—other nonhuman species) deserve to be accorded with specific and undeniable legal rights.

Wright quotes elephant researcher and National Geographic explorer Dr. Joyce Poole (with whom Katy Payne has collaborated):

> The range of their voices is astonishing, with some sounds produced by the larynx and others through the trunk. Many sounds that are well below the range of human hearing can be detected by elephants, sometimes more than six miles away. Sounds at such low frequencies transmit a replica signal through the ground, which means that elephants "hear" through their ears, their feet, and sometimes their trunks,

> too, recognizing the meaning of the call as well as
> the identity of the caller.[29]

Nevertheless, in June 2022, the US Court of Appeals ruled 5–2 that Happy was not entitled to the same rights as a human, specifically the right of bodily liberty as defined by *habeas corpus*. Both dissenting judges wrote about their decisions regarding the case: "Judge Rowan D. Wilson said the court had a duty 'to recognize Happy's right to petition for her liberty not just because she is a wild animal who is not meant to be caged and displayed, but because the rights we confer on others define who we are as a society.' Judge Jenny Rivera wrote that Happy was being 'held in an environment that is unnatural to her and that does not allow her to live her life as she was meant to: as a self-determinative, autonomous elephant in the wild.'"[30]

. . .

The more we realize that nonhuman species are conversing with each other creatively and with great complexity, the more it becomes necessary for us to listen responsibly. "Elephants, quite honestly, are running out of time," says Poole. "They're running out of space. And the numbers are declining. And they need more people to care about them."[31]

Payne's observations and insights in *Silent Thunder* are accompanied by tragic stories of "culling" in order to maintain herd size on ever-diminishing landscapes. Her empathic familiarity with individual matriarchs as well as complex generations of offspring within the elephant bond groups also means that she shares their profound grief over those who are killed. These deaths are not only caused by poachers for the theft of illegal tusks, but also occur as the result of official decisions to reduce elephant population in areas that cannot support their numbers.

. . .

The documentary *The Year Earth Changed* depicts footage from around the world during the first year of the COVID-19 pandemic. Massive decreases in human-caused noise, for one thing, meant dramatic reopening of space and tranquility from dense urban areas to vast oceanic regions (think: cruise ships). One website states it simply: "The film explores the impact of the coronavirus pandemic on wildlife and showcases the potential for human–wildlife coexistence to be achieved if humanity simply lightens its footprint on the natural world."[32]

Witnessing such restoration means renewed accountability for our claims to dominate and exploit nearly every inch of the planet—land and sea, wild and manicured, even including urban and suburban spaces occupied not too long ago by foraging species small and large. One story in *The Year Earth Changed* features the discovery of generational memory retained by deer who return to parks and fields within cities when human intrusion has been at least temporarily paused.

Another story reveals what one Indian village was able to achieve in order to live more harmoniously with a neighboring elephant herd once they realized—through a practice of deep listening and sensitive interpreting—that the group of murderous and crop-raiding elephants (who were being murdered in return) could be understood as needing a dedicated rice paddy of their own. Collaboratively selecting an area close to the edge of the jungle—which was clearly the herd's preferred location—villagers planted rice and offered it up, eliminating the competition between humans and elephants for food and space.

The solution seems so obvious, once appropriate mindfulness is engaged. Scarcity produces threats and violence; abundant sharing both creates and sustains peace. "When asked about his work to encourage local communities to work *with* rather than *against* wildlife, conservationist Dulu Bora says, 'We have lived in harmony with the elephants for years. I am just trying to remind them.'"[33]

. . .

Here is yet another important discovery: at least some elephants are apparently also listening to us. Bioacoustic recordings of elephants, called "ethograms," reveal their awareness of our noisy presence. Joyce Poole shares an example of her findings with NPR interviewer Lulu Garcia-Navarro:

> Let's say one member of the group will feel that they should go one direction and another member of the group feels they should go the other. They use something that I call a *Let's go* rumble. In other words, *Let's go this way. I want to go together*. But they may disagree on which way to go. So, you know, they'll talk back and forth and back and forth and negotiate. But often, these kinds of conversations go on when there are people in the area, and they need to decide where to go in relation to them. So, you can see them kind of listening to the sounds of people and having these discussions.[34]

What remains impressive about many of these researchers is that, despite their decades of listening practices, they are also humble about the limits of our ability to *fully* understand and interpret what we hear. Lawrence Wright refers to Poole's memoir, in which she describes a moment when, returning to an elephant camp after a three-year absence and accompanied by her infant daughter, she recognized "an intense greeting ceremony usually reserved only for family and bond group members who have been separated for a long time."[35] Nevertheless, Poole willingly acknowledges uncertainty: "Who can know what goes on in the hearts and minds of elephants but the elephants themselves?"[36]

. . .

The emotional content of so many narratives offered by whale and elephant studies inspires my hope that we may yet become partners in a project of universal reconciliation, human and nonhuman alike. As Dr. Rachel Naomi Remen says, we are surrounded by stories that have no ending because the listener is also adding to the story as it's being absorbed. In referencing the creation story found among so many origin myths and religious texts, Remen reflects upon the metaphor of an "accident" causing the world to explode into fragments that we are all together trying to restore into wholeness.[37] She speaks of the interactive healing power of telling/listening, not only in the context of medicine but also in the particular dailiness of connecting.

"How would I live," Remen encourages us to ask ourselves, "if I was exactly what's needed to heal the world?"[38]

THE SOUNDS OF LOVE AND WAR

"In Japanese, there is no way to say 'I love you,'" writes Nina Li Coomes. "Language is a labor of love. To understand each other, to listen, to speak with intention and precision is love in action, no matter if it is across two separate languages or one shared one. To tell someone you love them in a way they will understand requires thought and meaning. To be open and receptive to expressions of love requires similar thought and meaning."[1]

My German-born father neither spoke nor understood my mother's two first languages, Polish and Russian. Although my mother was a gifted linguist who had managed to absorb a lot of German as if by magic, the language my parents used for love was Swedish, which they had both learned as refugees in their late teens. My father carefully avoided speaking his first language except under very special circumstances. While I was growing up in our small town of Schenectady, English was the common family language. But when she spoke with her mother or with fellow immigrants from Eastern Europe, my mother almost always used one or the other of her two mother tongues.

In school, my sister and brother studied French while I studied Spanish (additional languages my polyglot mother more or less taught herself to speak). And yet, none of us mastered more than a couple of words in Polish or Russian. As for my father's German?

Auch nicht. We were expressly forbidden to study it in school. It was, after all, the *language of the murderers.*

It wasn't until, as an adult, starting a romance in Mexico with a Spanish-speaking man, that I found myself wondering how well my parents could have understood each other when they first met. Many multilingual speakers and academic linguists have suggested that we present as different selves in different languages, and I soon became conscious of what they meant. I am a bit more playful in Spanish because I hear myself sounding like a child; I become a person making silly verbal mistakes and at times not even making logical sense. When I can't find a specific word I need, I come up with a close-enough-for-me approximation. Although I've been told that my accent is excellent, I'm convinced that my conversational skills are mediocre. I can tell a few jokes and some brief anecdotes, but nuances of philosophy and politics are beyond reach, as are most articulations of complicated feeling. Even my own ears detect the limits of my capacity.

V. has an audibly Latino-tinged pronunciation when he speaks English; however, his vocabulary is far more substantial than mine is in Spanish. He reads literature in my language, yet to this day I haven't gathered the courage to try the same in his. I sense that V. has always been more confident than I am about being bilingual.

At the start of our relationship in Central Mexico, we agreed to speak primarily in English, though I could tell right away that we were both straining to speak in our authentic voices. From his halting rhythms, I knew V. was struggling to express himself fully in my language, and I was (perhaps even unconsciously) simplifying my own expressions to make sure he understood me. Later, when we were communicating via email long-distance, I typed, "Say it to me in Spanish?" I wanted him to write freely, without the extra effort for my sake. Immediately I saw not only the elegance of his sentences but also just how inadequate my own comprehension could be when faced with more sophisticated syntax and idiom. Surrendering, I copied and pasted his paragraphs into an online translation website. Even with the oddities of a computer-generated interpretation, the beauty of his writing stunned me.

Had I ever known this person? Did he know me?

The relationship between us ended after I decided we weren't compatible enough to sustain something long term, and I can't say this realization was unrelated to what I considered to be our language difficulties. In retrospect, I see that on multiple levels I felt haunted by my parents' long yet deeply unhappy marriage. What may have been a genuine ribbon of love between them often seemed hidden below their very vocal resentments and disappointments.

I think farther back to M., the boyfriend I had when I was an undergraduate, the one who worked as a counselor in a halfway house for teenage runaways while attending a private graduate school near my university in Palo Alto. He told me once that I didn't know how to talk to the kids "in their own language" and that I sounded like a "snob" because I was using a "fancy Stanford vocabulary." His criticism both stung and confused me. Was it wrong that I had only one voice? That it didn't occur to me to try to sound like someone other than myself? Having grown up in working-class Queens, M. was the first of his family to attend college. He was working on a PhD, but even before I heard a term for what he was doing, I could tell that he knew how to code-switch. And I apparently did not.

. . .

Is there code-switching in the animal world? "Marine scientists have observed that the melodies that orcas and humpback whales send through the sea have been changing rapidly over the last four decades. Instead of the psychedelic cetacean blues rhythms of the 1960s, nowadays the broken syncopes of grunge and rap pass through the waves as if the animals had adapted to the augmented underwater din" created by ever-growing numbers of cruise ships and container ships.[2]

Research shows a variety of ways that dolphin behavior changes to compensate for anthropogenic noise and its impacts on their echolocation practices. In one recent study, for instance, "the cetaceans turned their bodies toward each other and paid greater attention to each other's location. At times, they nearly doubled the length of their calls and amplified their whistles, in a sense shouting, to be heard above cacophonies."[3]

I'm here, I'm here.

Much worse than evidence of dolphins shouting is accumulated data that suggests our sonic booming (especially from military testing) is causing acoustic trauma to marine mammals. Short-term as well as long-term impacts can include decreases in feeding, which lead to delays in sexual maturation, increased infant mortality, and shortened life spans. Studies continue to measure (albeit with logistical difficulty) the associations between mass strandings of certain whale species and the use of sonars.[4]

. . .

Remarkably enough, there has been a dramatic resurgence in humpback whale populations off the eastern coast of Australia, thanks to the successful bans on commercial whaling. New studies suggest that there is less singing as a result of increased density in the male population; instead of using song to attract mates, these more abundant males are competing with "alternative mating tactics, such as mate guarding, surreptitious sneaker, territoriality, fighting, and displaying to gain access to a female."[5] This reduction in whale-song is interpreted by researchers as a form of behavioral plasticity in response to a social landscape where the risk of extinction has been minimized. "Singing was the more successful tactic in earlier post-whaling years, whereas non-singing behavior was the more successful tactic in later years."

Associate professor of biological sciences at the University of Queensland Dr. Rebecca Dunlop speculates: "If competition is fierce, the last thing the male wants to do is advertise that there is a female in the area, because it might attract other males which could out-compete the singer for the female . . . By switching to non-singing behavior, males may be less likely to attract competition and more likely to keep the female."[6]

. . .

Carl Safina has written that the singing of humpback whales was "entirely unknown to humans until the 1950s. US military personnel who'd begun listening for Russian submarines were astonished to realize that the strange sounds they were hearing were coming from *whales*."[7]

Although I suspect that there must have been generations of Indigenous fishermen and islanders who knew much more about this music than those in the so-called modern Western world, it seems ironic that military eavesdropping would be the source of such discoveries. As news of these recorded songs began to spread, our sense of interconnectedness with marine mammals promised more enlightened practices of environmental protections. Much like the

Gaia consciousness that developed in response to the first full view of Earth from outer space in 1968 (inspiring another "new" conception of the planet that Indigenous people had already held for thousands of years), it must have seemed as if humans were on the verge of recognizing the beautiful fragility of our shared home.

. . .

Nina Li Coomes writes that having more than one language in which to express herself is "holding two tongues in my one mouth." She uses the example of "the ability to carry sorrow and joy at once. Feeling it all together, allowing sorrow to co-mingle with fond remembrance, is what it means to feel setsunai . . . It reminds me of how I feel when I know there is a word for something in my other tongue; it reminds me of the pleasure of precise articulation and the frustration of not being able to bridge the gap."[8]

Poet Victoria Chang, in her book *Dear Memory*, writes: "Sometimes I wonder how much grammar my parents didn't pass on to me. On the other hand, I can speak another language, Mandarin, decently. I wonder what it would have been like to grow up in a family where everyone spoke the same language. The only language we had wholly in common was silence. Growing up, I held a tin can to my ear and the string crossed oceans."[9]

. . .

Dr. Viorica Marian reminds me that "several overlapping areas of the brain are involved in processing language," and research suggests that "decision-making in a second language tends to be more objective and more rational, less emotional."[10] My friend Bonnie says her mother is funnier in Cantonese and more practical in English. We are talking about the ways it's possible or maybe even inevitable to possess multiple personalities in a pair (or more) of languages. Her husband has learned how to say "Those are nice fucking shoes!" in several languages, including Cantonese (though he leaves out the curse word because he likes to say the sentence to Bonnie's grandmother, who still finds it hilarious every time). Their two young sons use the Cantonese word for "slippers," and in fact it's the only word they use for slippers; it's their family word. "*Taw-hi*," Bonnie texts me, with musical notes inserted to show me the word with its rising and falling sounds, helping me remember the way she sang it to me when we were having the conversation together in person.

This makes me think of my father always asking for *senap* in Swedish at the kitchen table (never *mustard*), and how my siblings and I used to mispronounce his word *schlafanzug* (pajamas) to say *schlaffen-suit.* It was one of the very few German words we ever heard him use with us on purpose.

. . .

Fierce Grace is a documentary about the life of spiritual teacher Ram Dass, with a major portion of the film depicting what he calls his *next incarnation* after having suffered a severe stroke. In one segment, his speech therapist explains that we can find new neural pathways for expression when our vocabulary isn't adequate or immediately accessible. Ram Dass is trying to describe to her what it feels like to be unable to retrieve certain words.

"The clothes," he begins, "are in the closet . . . but I can't get them out."

The speech therapist seems delighted by what she has heard, insisting that Ram Dass has solved his own puzzle by creating a metaphor. This is human neuroplasticity in action, the brain rewriting its territory. Such inventiveness can be another demonstration of how we can stretch our voices and our ears, how sometimes we can be more poetic in our less-fluent tongue.

. . .

When my Sweden-born Aunt Elsa and Sweden-fostered Uncle Eli visited during my childhood, I always begged them to speak Swedish with my parents. The melodious sentences were so much more beautiful to me than their English counterparts. For a while, at least, I didn't care that the conversation excluded me; it was the harmony I wanted. Maybe I understood this was the vocabulary of happiness, for all four of them, a score of romance and innocence and partnership of the heart.

No surprise that I wanted to visit Sweden in person, eventually, to learn at least a few of my own words, phrases, fragments of an extra self. I vaguely wondered if I'd fall in love there too, but mostly what fascinated me was the idea that I might have been born there, had my parents chosen to stay. I might have become a variation of myself but in another voice and landscape, with an entirely parallel identity, or not.

I was twenty-two when I made my first visit to Sweden, staying with my Aunt Elsa's sister in Gothenburg for a few months while working as a dishwasher in the family's restaurant, one that specialized in crêpes. For a long weekend, I took a train up to

Stockholm to meet the two couples who had been my parents' closest friends immediately after the war. One said matter-of-factly that my father had been lucky for "getting out in time" and "not being sent to a camp."

Shocked, I insisted that he had spent the last year of the war in Buchenwald. At first, they disagreed with me, said I was wrong, that they would have known.

"But I'm his daughter," I said. "And I know for sure."

At that point we all sat quietly for several moments around the dining table. I wrestled with the idea that he hadn't told them about being in Buchenwald, or that they had forgotten. Neither seemed probable or even sensible to me, except when I imagined that maybe it had all been so recent he couldn't find adequate words for what he had lived through. Nevertheless, I couldn't stop thinking about the black-and-white photos I'd seen, images of him with my mother, clearly in love and posing together somewhere on the Swedish coast in the dazzling brightness of summer. Despite his smile, it was the slenderness of his ankles—the voice of his body—which seemed to be telling the backstory.

Meanwhile, although I tried my best to learn conversational Swedish, it seemed that my accent was problematic. On the phone, when I risked a few words, my father insisted I was speaking the regional dialect of *Göteborgska*, not *Stockholmska* as he and my mother had learned. Both of my parents seemed to be mocking me, perhaps subtly or overtly discouraging me from immersing any further in that place they had left behind, even though it had been a temporary refuge. Maybe this was yet another way of warning me not to stray too far from their secure landing site in the United States.

I experimented with a Swedish boyfriend anyway, tried enduring the grim palette of winter, reveled in the long days and brief nights of Scandinavian summer. A part of me loved it there, but I couldn't stay. I was a stranger in a strange land, maybe a variation of the way my parents once felt, never quite allowed to forget we had come from elsewhere.

. . .

In a 1997 video of my mother giving her Holocaust-survivor testimony for the USC-based Shoah Foundation, she is asked to talk about her earliest memories from her hometown of Vilna before the war. My mother describes riding a horse-drawn sleigh through the snow to her grandmother's house in the Polish countryside, while wrapped in fur blankets and listening to the tinkling of bells on the horse's halter and reins.

"Like something out of Pushkin," she says, her face illuminated and softened with nostalgia. But the person behind the camera must have revealed a blank expression at the mention of this Russian author's name.

"You don't know *Pushkin*?" my mother exclaims directly into the camera, as if speaking directly to me, to all of us in the future. She shakes her head with an obvious mixture of shock and pity. "Pushkin was my *guy*," she insists.

I think of the well-worn books lining our childhood shelves, those titles in the mysterious Russian alphabet, inscrutable to the entire family except for my mother. *You refused to listen.*

Minutes later, describing her arrival as a refugee in Sweden with her parents in 1947, she admits on camera that at first, "the language was impossible. For three months I just listened to the melodies, trying to make sense." Waving her hands in the air, creating what looks like a river with rapids, she smiles again. "Then suddenly: it clicked."

In response to another question from the interviewer, my mother explains that although she was the one who insisted upon moving to America, she always regretted not remaining in Israel, where she and my father were married in 1951. Because I have no memory of hearing her express this particular regret, I struggle yet again to imagine who I might be now had I been born and raised in *that* country. In Israel I would have been required by law to serve for two years in the army. What alternate shape of me might have been created by the sounds of love and war in *that* world? And how many possible selves do we carry?

. . .

"We listen not only with our ears but also with our body," writes J. Martin Daughtry, author of the book *Listening to War: Sound, Music, Trauma, and Survival in Wartime Iraq*. "We flinch against loud sounds before the conscious brain begins to try to understand them."[11]

I never asked my parents if they were haunted inside their sleep by echoes of military parades or gunshots, if they ever compared notes about their recollected sounds of the war. My father's Shoah Foundation testimony includes only a single reference to an experience that he initially calls "quite frightening." In a measured voice, he describes the morning after *Kristallnacht*, in November 1938, when uniformed German authorities showed up at his Jewish school in Hamburg to terrorize the students; all the while, the synagogue next door burned to the ground.

"That was traumatizing," he concludes, his tone subdued and his eyes brimming.

My mother, in her Shoah interview, says briefly, "I remember the bombs," mentioning the first few months after she and her parents were liberated by the Russians from their hiding place. "I was afraid of the bombs."

. . .

In Anne Karpf's book *The Human Voice*, she cites the story (from Robert Fulghum's massively bestselling book, *All I Really Need to Know I Learned in Kindergarten*) about the South Pacific village in which the loggers, when faced with a tree too large for felling by an axe, select someone to scream at it loudly for thirty days. At which point, "the tree dies and falls over. The villagers claim that it works because screaming at living things kills their spirit."[12]

. . .

In contrast to many who are composing and promoting music as healing for any number of physical and/or psychological illnesses, Bernie Krause says this: "Anecdotal evidence strongly suggests that natural soundscapes and biophonies, in particular, may lower stress indicator levels (glucocorticoid enzyme, heart rate, blood pressure, and so on) in humans far more successfully than environments saturated with music, because music is culturally biased and may actually produce a result opposite from what was clinically intended or hoped for."[13]

. . .

It may not go without saying that musical preferences vary among generations and eras—and even more importantly—vary widely across cultures. Author Garret Keizer, in his book *The Unwanted Sound of Everything We Want*, explains that "the essential difference between music and noise is neither acoustic nor aesthetic but ethical."[14] Arguing that noise and volume control are essentially a form of sociopolitical power, he asserts that what matters is the ability to force sound upon someone who cannot walk away.

. . .

It has been asserted by numerous World War II historians, political scientists, and philosophers that the amplification of Hitler's voice by means of loudspeakers played a significant role in the near-hypnotic behavior of an entire nation, not only soldiers but civilians including women and children. Worshipful crowds of tens of thousands, hundreds of thousands, were captured on film: a sea of uplifted faces and arms, a wild chorus of their cheers providing us with indisputable evidence of their unified ecstasy.

Originally invented by a Danish man named Peter Laurids Jensen, the Magnavox "was first used to play opera to a huge happy crowd in San Francisco."[15] Further developed during World War I in Germany (by the Siemens company), large sound amplification systems were designed for public spaces and political gatherings; by 1932 they were being used to great effect in the staging and orchestration of Nazi rallies. Modifications to the size and placement of speakers throughout a very large area were made so that in the delivery of Third Reich propaganda with a mile-deep audience, "perfect synchronization" could be achieved "between the masses and their leaders."[16]

"Hitler's goal was to let his voice be heard, quite literally, by as many people as possible." Within a few weeks in 1932, "in the course of 200 mass events taking place across the Reich, the party addressed over ten million people."[17] As an unprecedented form of acoustic dominance and control of listening, no one could "escape" the "sheer brainwashing loudness of Hitler's rants."[18]

In an online article about the military history of the loudspeaker, its advantages are listed in the following manner: "1) It allows for immediate exploitation of a target audience in a fluid battle zone; 2) It overcomes illiteracy; 3) Operators can be easily trained; and, 4) It is impossible for enemy leaders to prevent their soldiers from hearing the broadcasts."[19]

For the Third Reich, technological modifications created so much resonant feedback that millions of marchers at rallies and parades were collectively listening to themselves and each other as they massively echoed and marched.[20]

. . .

After the death of his father, when I was thirteen, the few times I heard my father speaking German were when he spoke on the phone, long distance, with his brother Joseph in Israel. Even now, I imagine that between the two of them, *the language of the murderers* remained instead the language of their childhood in Hamburg, the vocabulary of their shared survival in Buchenwald, a safety zone that needed no translation.

. . .

"Is language a place you can leave?" asks poet Ilya Kaminsky. "Is language a wall you can cross? What is on the other side of that wall?"[21]

He is referring to what he calls "the language of war" and wrestling between Ukrainian and Russian. Once again, I wish my mother were still alive so I could ask her to help me understand more about her passionate love for the Russian language, Russian literature in particular. What was it like for her not to share Pushkin's poetry with the man she was married to for fifty years? What was it like to live in a country for whom, throughout the Cold War, Russia was the enemy?

In February 2022, Vladimir Putin's murderous dictatorship invaded the sovereign nation of Ukraine, raining bombs on civilians. In addition to a national boundary, many in the victimized population share a primary language with their killers.

. . .

In a now-famous study of "experimental generation of interpersonal closeness" published in 1997, a group of five social psychologists attempted to induce a temporary feeling of interconnectedness and intimacy between paired individuals who spent about an hour asking each other personal questions. Selected to have a basic compatibility by virtue of a brief questionnaire regarding attitudes and attachment styles, the paired subjects reciprocally answered a total of thirty-six questions designed around "self-disclosure and relationship-building tasks that gradually escalate in intensity."[22] One reason the study became so famous is that a writer named Mandy Len Catron referenced its results in her Modern Love essay in *The New York Times* in 2015. She was especially intrigued by what she called the study's "most tantalizing detail: Six months later, two participants were married. They invited the entire lab to the ceremony."[23]

Catron invited a friend of hers to try the experiment; while she liked learning about herself through the answers, she "liked learning about him even more." Reflecting further on the process, which concludes with four minutes of staring into each other's eyes, Catron wrote: "I felt brave, and in a state of wonder. Part of that wonder was at my own vulnerability and part was the weird kind of wonder you get from saying a word over and over until it loses its meaning and becomes what it actually is: an assemblage of sounds."

. . .

My friend Luisa tells me that in her childhood home she learned both Italian and Sicilian. She could always tell when her mother and grandmother were speaking Sicilian so as not to include her father who only understood Italian. In Sicily, Luisa explained, people can tell exactly which part of the island you are from not only because of the way you pronounce certain words but also because of the words you use to name one thing or another, such as a certain type of pastry.

Variations in regional accents and idiomatic expressions are of course common throughout the world, but it's intriguing that even a relatively small island can be divided into parts, and that your listening brain will be trained to differentiate between the voices of your closest neighbors and those who are living at a slightly farther distance. Once oceans get involved, and then centuries, it's no wonder the nuanced resonance of tongues becomes even more significant.

Nancy Rocha, a multilingual interpreter, tells me that in Mexico, the words for "to hear" and "to listen" are basically interchangeable, while in Spain, there are important distinctions. Hearing is mechanical, an almost-automatic behavior, while listening requires deliberate and specific intention. She demonstrates this latter mode by leaning forward with her upper body, inclining her head at an angle, widening her eyes.

"I'm listening in Spain now," Nancy says. *Estoy escuchando.*

. . .

Bonnie told me about the "marinating soup" of Cantonese as her language of home; up to the age of five, it was the only language she spoke, until she went to kindergarten and began to speak English.

"Your mother joined your father on *his* island," Bonnie said, when I mentioned that my father didn't learn his wife's languages, even though she learned German. I think about Mom being so animated whenever she spoke Russian with friends, how her extra-dramatic self was illuminated and on display, how those Russians who knew her told me after she died that she spoke an exceptionally poetic Russian, a language already poetic. And I think of words that can be avoided or steered around—an entire language even, renounced but not forgotten.

. . .

"There's an expression in French, *avoir le cul entre deux chaises*, which means, literally, sitting on half of one chair and half of another," writes David Hoon Kim, describing some aspects of his complicated bilingual existence.[24] It's a telling image: the embodiment of a precarious and uncomfortable position, a status that is physically unsustainable for very long.

This makes me envision my family home as a Tower of Babel, cacophonous with languages and misunderstandings, chasms I kept trying to navigate like the dark hallway I crossed in an effort to escape from my nightmares. As pediatrician and psychotherapist Donald Winnicott famously said, "It is a joy to be hidden and disaster not to be found." Paradoxically, my longing to be seen and understood—a desire shared by children as well as adults—was also complicated by my need to retain a secret self, a space in need of protection from invaders, and even, at times, protection from too much noise.

. . .

Although it's considered a harmless condition, Exploding Head Syndrome (EHS) is a rare sleep disorder that is "defined by episodes that typically occur during the transition period between sleep and wakefulness. These episodes feature imagined sounds or sensations that create the perception of a loud explosion and possibly a flash of light, in the sleeper's head. The episodes are brief, usually lasting less than a second . . . Primary management of the disorder includes education and reassurances about its benign nature. Some people even experience fewer EHS episodes after hearing this information."[25]

Thai filmmaker Apichatpong Weerasethakul, who suffers from

EHS, has written that when he talked to his therapist about the hallucination of sounds of gunshots resounding in his skull, she told him that "maybe the sound came from the veins behind my ears, that maybe it was an internal pressure before dawn. I thought there was a symptom called 'ghost ears' or maybe I was possessed by the sounds of the past."[26]

. . .

"In the Mebêngôkre language—spoken by the peoples who call themselves by this name but are known in the white world as Kayapó and Xikrin—it is common for speakers to use the expression 'Gamá?' when talking over the radio. This translates something like: 'Was your ear able to hear to understand?' The interlocutor then replies: 'Arup ba kumá.' Which means: 'Yes, I was able to understand what I heard in my ear.'"[27]

In an essay about Indigenous peoples of Brazil faced with near-extinction due to impacts from government-built dams, journalist Eliane Brum writes with audacious self-consciousness about her role as an *escutadiera*, or listener. "I try to understand what I hear with my ear. And with all my other senses, intuition included. I suspect this is what 'ear' means in the Mebêngôkre tongue—it is a more-than-body-part. When I listen to the Indigenous, 'hear in my ear' is about hearing what I understand them to say, when they say it in my own tongue. But what tells me more is what I can grasp from their words when they are spoken in a tongue I don't understand."[28]

. . .

Occasionally while walking in my neighborhood in North Berkeley, I can't help overhearing the sounds of Spanish on my street. After returning from an extended visit to Mexico, I'm extra attuned to the casual conversations among Latino laborers, the ones perched high on rooftops and talking on their cell phones inside trucks whose windows are rolled down to let in the warm air. They don't know I understand what they're saying, or maybe they don't care; maybe I'm invisible to them the way they are invisible to most others, even those who hire them to build stone walls for their gardens, or to rip off the old shingles and replace them with new ones.

I'm not adept enough to discern whether the speaker might come from Mexico or Central America or even points farther south.

One day a few years ago, a Peruvian gardener said to me, "You talk like a Mexican!" and I couldn't tell if he was making a joke or an observation. I thanked him. It felt like a compliment, because I find Mexican Spanish so enunciated and clear and comfortable.

Is it my melody? My rhythm? My accent?

. . .

When, during a Zoom interview, I ask where she thinks her own third ear might be located, Anna Deavere Smith tips her head slightly to one side, as if she's allowing the words to enter her through some unseen portal. Anna doesn't point anywhere—but knowing we are both dog lovers, I mention the impression she has created of a dog cocking its head to listen. We both smile. *Third ear.*

Not between the two but somehow beyond, greater than the sum of our parts.

Maybe my parents didn't just meet in a third landscape, a third country, or third language—but with a commitment to reading between the lines, listening to each other through their hearts. Maybe they whispered with porousness, a semipermeable alphabet, a form of refuge that sounded like hope.

HUNGRY LISTENERS

"Everyone's a history," says Whit Missildine. You could call him a professional listener, although I haven't asked how he might feel about that designation. Whit is the host of a popular weekly podcast called *This Is Actually Happening*, a show centered on first-person stories of people who have been through what he calls "a massively life-changing event that was destabilizing—something difficult to make sense of." On the show's website, it's described as featuring "uncanny, extraordinary, true stories of events that have dramatically altered the lives of ordinary people told by the people who lived them, exploring the question: What happens when everything changes?"[1]

Because we live in the same neighborhood, Whit and I meet at my house. We are seated on opposite ends of a couch in my living room, oak trees and sky beckoning from windows, intermittent birdsong providing commentary in the background. Whit acknowledges that this is going to be a rare moment in which he is the subject, so I decide to start in his comfort zone. What's it like when he does the asking?

After conversing with his selected subject for hours, Whit carefully ("and heavily," he says with a grin) edits himself out of the finished piece. Unlike most other podcasts, which include a reporter or interviewer—"intervening too much," according to Whit—he is deliberately replacing his role as listener with the receptive presence of the podcast listeners. The words of the (often-anonymous) storyteller are the only ones we hear, and his or her narrative is "like a voice in your head."[2]

Whit firmly believes that "there is a universal need for deep storytelling," and with this he is referring to all aspects of the series he has been creating since 2008. "People have a hunger to tell their story," he says. If the record numbers of podcast subscribers are any measure, it's clear that there is a rapidly growing audience of people who are even hungrier to listen.

. . .

"Podcasting is a peculiarly intimate medium," writes Rebecca Mead, in her *New Yorker* article from 2018 on what she calls the "new golden age of storytelling."[3] Mead explains that because these audio narratives are typically received in a solitary environment by way of headphones or while driving alone in a car, the experience "can be immersive in a way that a radio playing in the background in the kitchen rarely is." She also cites the study by Edison Research (a group that people like Whit Missildine follow assiduously in order to track current listening trends) "which found that nearly a quarter of Americans listen to podcasts at least once a month."

More recent statistics are stunning evidence of growth during the past few years; as of 2023, there are "464.7 million podcast listeners globally . . . and it is predicted that there will be around 504.9 million podcast listeners worldwide by the end of 2024."[4]

. . .

Aside from whatever personal strain might be involved in the sharing and recording of intimate narratives, not to mention the labor of the editing process, listening to an extended monologue like the ones in Whit's podcast takes effort, concentration, and intensity of focus that is distinct from automatic and habitual listening. What are we searching for?

According to various studies, most people who regularly subscribe to long-form storytelling podcasts are selecting them in order to be entertained, informed, challenged, and occasionally overwhelmed by their own feelings or memories. Even now that we are mostly past the COVID-related restrictions that resulted in large numbers of people living and working in relative isolation, it also seems increasingly clear that narrative podcasts are providing us with a form of companionship as a balm for our loneliness. Whether cooking or stretching, half-distracted or intently absorbed, we are no longer just choosing music to add soundtracks to our lives. We are deliberately saying *Yes* to these voices that are whispering, bragging, explaining, and confessing directly into our ears.

. . .

The backstory for *This Is Actually Happening* begins with Whit Missildine's graduate work at an HIV treatment and support center in New York City during the years between 2002 and 2005. He was participating in a study of what is called "motivational interviewing," in which the interviewers do not limit themselves to asking questions but extend into offering guidance and advice as part of the conversation. Due to his youth and inexperience, Whit was designated as one of the "control" interviewers whose purpose was to remain entirely neutral while interacting with his subjects. Knowing that he was collecting personal narratives that would be turned into "data," Whit realized by the end of the study that he "felt an unmet need to tell their raw story."

Why? What can explain this longing to be tuned in to someone else's existential crisis? "Things aren't what they seem," says Whit. "I think that my listeners are people who are aware of their mortality, who have been through something themselves. Maybe an invisible illness, maybe a trauma of some kind. I don't know why exactly—but I know I feel liberated by these stories."

As Whit and I drink cups of tea, I ask him if he might share more about the personal origin for his podcast. In his early thirties, not long after his participation in that HIV interviewing study, Whit suffered a shattering period of what was eventually diagnosed as a panic disorder.

"Have you ever been through something like that?" he asks me.

I tell him that although I recognize some of the symptoms, I haven't felt anything nearly as acute as his descriptions of feeling as though he was dying.

"So, of course I was personally fascinated by other 'crazy stories,'" he says, shrugging. "People who had been through something complex."

He acknowledges having been inspired by oral historian Studs Terkel and some of the short documentary films made by artist Andy Warhol, both of whom depended on intricate collaboration with their self-revealing subjects. During encounters with the curated narrators for Whit's show (people he used to find via Craigslist, where he offered $10 or $20 per story), he asks probing questions, prompting them to dive more deeply into the emotional content.

"I'm not all that interested in where it happened or when, those kinds of facts," he says, helping further define what sets his interviews apart from what might be considered a more traditional approach. "I'm much more interested in finding out what it *felt* like, what they were experiencing on the inside."

Nowadays, Whit's website provides a basic form for submissions, in which potential storytellers sketch a very brief description of the event(s) they want to tell about. Once storytellers are chosen, Whit actively helps to shape each teller's narrative by asking questions in a three-part approach: 1) background/context; 2) what happened; and 3) the aftermath/reflection.

"Sometimes there is very little background," Whit says, "and they dive right in." What surprises me most is when Whit mentions that he doesn't care whether or not the stories he records are entirely true. "My goal is to help build compassion and empathy," he explains. "I think my podcast listeners are getting to relax their own egos, to deepen their curiosity about what it's like to be another person."

With over three hundred episodes stretching by now across thirteen seasons, *This Is Actually Happening* was recently ranked number twenty-two on *The Atlantic*'s Best Factual Podcasts. With a click of a button, you can find monologue after monologue featuring forty-five minutes of people's most extreme life moments, the kind I might call thresholds of Before/After.

The titles of the episodes are always framed as "What if . . ." ("What if you bled to death?" "What if the devil was battling for your mind?" "What if you faced the monster in the mirror?" "What if you went missing?" "What if you grew up in a cult?" "What if you survived a genocide?" "What if you were a human shield?"). I admit to being equal parts fascinated and horrified by the sheer volume and tone of these labels, which often strike a chord somewhere between carnival sideshows and reality television. Whit can't know if his listeners are being turned on by some massive quotient of *schadenfreude*. But it's equally possible they are experiencing vicarious terror as well as transformation.

Whether or not you have ever derived some measure of joy or even mild gratification from observing and/or listening to the struggles of others (i.e., the very definition of *schadenfreude*) it seems increasingly clear that humans may also be wired for the opposite, something called *freudenfreude*. Joy in the successes of others. Researchers at places like the Greater Good Science Center in Berkeley, California, are busily searching for (and finding) evidence that cultivating supportively positive networks among friends, colleagues, even strangers—celebrating stories of "wins" and cheering for good news, boosting teamwork and empathy—can lead to personal as well as collective well-being.

Many of the latest studies in the category known as "self-help" are showing that we can learn about (and remember) ways to practice as well as amplify our own happiness, kindness, and resilience—and to thereby help the happiness, kindness, and resilience of others. Connection is not only an outcome of such activated listening behaviors, but also a source of even "greater good," discoverable both inside and out.

. . .

Soul Music is a BBC Radio 4 podcast series about "music with a powerful emotional impact" that has been on the air since the year

2000. Each thirty-minute episode revolves around a single piece of music, ranging from iconic classical pieces like Puccini's "Nessun Dorma" and Beethoven's Fifth to modern legends like Amy Winehouse's "Back to Black" and Prince's "Purple Rain." Lacking any introductory voice or host, an episode of *Soul Music* presents three or four distinct narrative segments, in which each person attempts to articulate a particularly meaningful relationship to the identified selection. The show was the subject of a recent article in *The New Yorker* by Hua Hsu, who writes, "Listening to the show can be a dreamlike experience, and it sometimes feels as though the voices were in conversation with one another."[5]

Rather than choosing popular songs and then trying to find a variety of commentators to create a sequence, the relatively small podcast team moves in the opposite direction, "scour[ing] forums, message boards, and blogs for people's stories, looking for surprising resonances . . . The show's brilliance lies in the power of people trying to explain the flood of memories that a song triggers, and in the realization that this is always happening, everywhere."

The idea is to hear people narrate their own otherwise-private listening practices, while the podcast allows the musical selection itself to float nearby, simultaneously background and foreground. Hsu describes the podcast as "diaristic and slow, as people search for the right words to describe the moments of beauty or sorrow that a song evokes." The recorded voices commenting are only occasionally those of musicologists or so-called expert musicians; more often they are what might be considered ordinary individuals of all ages and backgrounds who share a deeply personal sense of connection to the piece in question.

After scanning the list of hundreds of episodes, I randomly click on one about "Killing Me Softly with His Song," and I'm startled to recognize, by sheer coincidence, that the first voice belongs to a friend of mine named Julie Daley. She is reflecting decades back to a memory of listening to the radio in the car at age sixteen with her boyfriend; she married him after they became pregnant. The next voice comes from a musicologist talking about the song's chord structure; after that comes a segment on the origin of the song's creation by its writer, Lori Lieberman, whose lyrics were inspired by listening to singer Don McLean at a concert. Someone else tells the story of Roberta Flack's Grammy-winning cover version from 1973, followed by more covers, including one by the Fugees that won another Grammy in 1997. By the end of the half hour, the episode returns to its first narrator, my friend Julie, for whom the song eventually became associated with death as well as a determined restoration of possibility.

Although I've known Julie for ten years, and though I am familiar with a good part of her history, until now I've never heard the story of her falling in love and the death of her husband in exactly this way, circling itself around the bittersweet heart of "Killing Me Softly."

"The show's best moments involve wondrous feats of listening and imagining," Hsu explains. It's "about epiphanies, not nostalgia. What you're left with is a yearning for your own discoveries, set to your own songs . . . 'Soul Music' is less interested in telling us how to hear a song than it is in encouraging us to listen."

. . .

In Elizabeth Hellmuth Margulis's book *On Repeat: How Music Plays the Mind*, she ponders the question of "why it is we crave repetition when it comes to music, either in the catchiness of a beat or our joy at hearing the same songs or pieces over and over again. It is clearly a neurological mystery. Our brains and bodies must gain something from re-hearing the same assemblies of sound. It must bolster us, not bore us."[6]

. . .

Rivka Galchen, in her *New Yorker* article about the renovation of Geffen Hall in Manhattan, explains that "psychoacoustics is the study of how mood, color, sense of place, and other emotional factors affect the way people perceive and understand music." Referencing her conversations with the two principals of the company hired to improve the hall's acoustics, she mentions that although past experts in acoustics "relied primarily on what was easiest to measure—things like frequencies and reverberation times . . . this started to change in the nineteen-nineties, when acousticians 'began to rely more upon their ears, informed by historical precedence, than their measurement devices.'"[7]

By the end of the article, Galchen cites a scene from the very first of Leonard Bernstein's televised *Young People's Concerts*, a series that aired on CBS from 1958 to 1972. "Now we can really understand what the meaning of music is—it's the way it makes you feel when you hear it," Bernstein says. "You see, we can't always name the things we feel . . . and that's where music is so marvelous, because music names them for us, only in notes instead of in words."

From his first episode entitled "What Does Music Mean?" through a total of fifty-two others, including, for instance, those he called "The Sound of an Orchestra," "Forever Beethoven," and "Bach Transmogrified," Bernstein's final episode appeared on CBS in March 1972. It focused on British composer Gustav Holst's *The Planets*. In this program, the maestro explains to his audience that Pluto had not yet been discovered when Holst composed the piece, and even though, when Pluto was found and identified in 1930, Holst was still alive, the composer chose not to add this new planet to the piece. Yet Bernstein, along with his New York Philharmonic, announces his plan "to make up for this omission by supplying a little Pluto of our own." He goes on to say that this Pluto piece will be improvised, allowing the musicians to "be as surprised as you at the mysterious sounds we will be making. In other words, you are about to hear a piece nobody has ever heard, nor will ever hear again. It's a once-in-a-lifetime experience, a real spaced-out trip. And here it is: Pluto the Unpredictable."[8]

I wonder what Holst (who died in 1934) and Bernstein (who died in 1990) would make of the unpredictable redefinition that occurred in 2006, when the object formerly known as the planet Pluto would have its status downgraded to that of a dwarf planet because it did not meet the third of three criteria designated by the International Astronomical Union. In the decades since that 2006 resolution, however, debate has continued among astronomers around the world with regard to the agreed-upon requirements for planethood.[9]

. . .

A friend of mine once said to me, "I was an only child, always hanging out with adults and adult conversation. I had big ears."

Anna Deavere Smith writes in her book *Talk to Me: Listening Between the Lines* that Eudora Welty said about her own childhood, "She would sit in the hallway outside the room where all the adults were talking, and her ears 'would open up like morning glories.'" Smith is explaining her own practice of listening very closely to the melodies and rhythms of people with whom she is conversing, recognizing a particularly nuanced shift that signifies a vulnerability or revelation of some kind, when "the music of the moment overpowers the information they are trying to communicate."[10]

. . .

Every once in a while, I imagine I can hear my heart's soft, persistent thrumming, the valves opening and closing, harmonizing and syncopating, blood pulsing as it rushes into two chambers and rushes out the other two, again and again. How brilliant is the design of the body—meaning, my heart knows what to do without my telling it what to do. My veins and arteries do their work silently and continuously, managing oxygen and hemoglobin levels, coordinating nourishment along with my immune system. Semipermeable membranes and mitochondria and cell division. What language do they speak to each other within the boundaries of my skin and my bones? What does my heart want me to know?

. . .

In science journalist Florence Williams' newest book, *Heartbreak*, she examines the idea that our "cells are listening" to our emotional experiences.[11] According to her research, our suffering (such as the devastation following a divorce like her own) impacts our organs, especially our hearts. Our blood can reveal how sad we are, how much we are brokenhearted. This is *not* a metaphor.

After twenty-five years of marriage and its abrupt, unexpected end, Williams undertakes a series of journeys combining professional and personal curiosity; in seeking to understand how loss and grief might be transformed, she heads into the wilderness for a reckoning with herself.

Williams's previous book was entitled *The Nature Fix: Why Nature Makes Us Happier, Healthier, and More Creative*. Modern research on well-being keeps scrambling to catch up with the oldest understanding of what humans need in order to thrive; as Williams herself puts it: "This book explores the science behind what poets and philosophers have known for eons: place matters."[12] Ironically, despite evidence that habitat preferences—including, at the very least, easy proximity to natural spaces—are widely agreed-upon among many different cultures and eras, Williams quotes a fellow researcher who says that "studying the impact of the natural world on the brain is actually a scandalously new idea."[13]

. . .

It makes perfect sense to note that Beethoven is one of many composers who have "dedicated symphonies to landscapes . . . [that] our nervous systems are built to resonate with set points derived from the natural world."[14] Given that loneliness is measurably hard on our immune systems, it helps to be reminded that healing connections to repair a broken heart aren't only to be found in new love but also can be felt in the awe of mountains and the music of rivers. Maybe what renews our life force is the soundscape of an underwater song, a rumble sensed through our skin, through the soles of our feet, or through our many variations on a third ear.

Let's go this way.

I'm here. I'm here.

. . .

Near the end of my interview with Whit Missildine, we talked about the cultural pressures to be happy and to "avoid suffering, as if it's something we can actually control." We spoke for a few minutes about the Buddhist value of accepting what is. And we agreed that in listening to the often unresolved struggles of another human being, we get to feel, at least for a while, a little less alone.

"What if they couldn't wake me up?" In season thirteen, episode 308, Whit became a storyteller on his own show, just two weeks after having collapsed into a non-responsive state that lasted about five hours. In the still-recent aftermath, he explains the resonance and comfort he summoned from previous storytellers on his show, whose own experiences provide him with a "golden road map" to navigate a series of awakenings and feelings ranging from euphoria to profound loneliness.

Almost five years past our conversation in my living room, I hear Whit echo some of his previous thoughts about interconnectedness. "When we listen to each other, when we honor our own story and the story of others, when we do so with love, the unknown becomes bearable. We can walk through it together."

In the podcast, Whit acknowledges that the show "asks so much of listeners. It asks you to sit not only with someone else's pain but with someone else's complication and contradiction, their shadows and their light." And a few moments later, his voice breaks with emotion. "I really carry these stories in my heart." Whit takes a shaky breath before continuing. "There are over 300 stories now, and all of them have been these teachers. I've known that over time, but I *know* that now. I can feel how these stories and all of you live in my system."[15]

THE HEARING TEST

WHALE EARWAX IS STORED IN LAYERS THAT ARE NOT unlike the rings of a tree. And it turns out that the stress hormone cortisol is found contained in these layers—which accumulate as hardened plugs for the entire lifetime of a whale. Comparative physiologist Stephen Trumble and a group of his colleagues from Baylor University studied twenty plugs from three baleen whale species from the Pacific and Atlantic oceans (humpbacks, fins, and blue whales). They determined that variations in what they termed "survivor stress" over a period of 150 years (1870–2016) could be directly correlated with human-induced factors—such as "exposure to the indirect effects of whaling, including ship noise, ship proximity and constant harassment."[1]

Although there was a dramatic decline in stress levels related to bans on commercial whaling in the 1970s, it appears that cetacean stress levels have been climbing again—especially more recently—alongside rising water temperatures. Trumble explains that scientists like him are still trying to "narrow down exactly what about climate change is causing the increase in stress." As he puts it: "These whales truly mirror their environment and can be used in a way similar to the canary in the coal mine."[2]

. . .

Who could have predicted that the year 2020, the year of perfect vision, would turn into the year of "We can't hear you," "Can you turn on your microphone?" and the ubiquitous "You're muted. Unmute yourself." All of our worry about not being able to hear each other has been amplified by how much harder it is to listen—even to ourselves—across all of this distance. Our ongoing separations are stretching far beyond the prescription for safety, that (obligatory-for-a-while) six feet of distance.

Many of us are painfully aware of the stresses caused by our isolation, our fears of infection and contagion, our debates about how and how much to protect ourselves as well as the most vulnerable among us. And though no one can yet say what the long-term effects of these physical, emotional, and social impacts will be, even in the short run, virtual rather than actual interactions are straining our nervous systems to their limits.

You could say that the wood-wide web has expanded us. The pandemic silenced us yet perhaps also gave us back to ourselves in a quieter, kinder way. And yet, notice how these words attempting to describe the moment create a haunting resonance of their own. Muting and unmuting; survivor stress; canaries in a coal mine. Despite what might be considered the best of intentions, these phrases, to my ear, neither sound nor feel like reassuring metaphors.

Listening feels more complicated than ever, and more urgent.

. . .

Midway through the year 2021 I woke up one morning and something was noticeably wrong with my right ear. It was as though I had inadvertently stuffed it with cotton. As a kid, I used to suffer from chronic ear infections, what they used to call "swimmer's ear," due to all the time I spent in the water. Since I hate wearing earplugs, I had to learn various tricks to make sure my ears weren't staying damp after a swim, in order to prevent bacterial infections.

On this particular morning, though, I knew I didn't have an infection, because nothing hurt; nothing was red or even sensitive to the touch. I just felt plugged up. Muffled. I kept trying to yawn as if I were on an airplane, kept blowing my nose one nostril at a time, trying to regulate the pressure or open up some inner space. The simple remedies I had practiced for decades. But nothing worked, and the sounds stayed strange, damped down.

The easiest solution presented itself: I would get my ears cleaned. At which point it would be obvious that the *problem* with my right one was simply caused by some extra wax buildup, and a washing process by a professional would do the trick. I'd be so fully restored that anything else—an audiogram, for instance—would be optional. Maybe just a bit of research, to satisfy my curiosity about a baseline measurement to have on hand for possible future loss.

. . .

When I call to schedule an appointment with an audiologist in December 2021, I mention that I want my ears cleaned—but might also be interested in a hearing test because I have noticed some reduced clarity in one ear.

"Oh, you sound too young to worry about that," the man on the phone says.

"I'm almost sixty-two," I say.

"Spoke too soon!" he self-corrects. "That's just about the age we do notice some loss. Let's see what we find when you come in."

Two days later, as soon as I arrive, I take an immediate shine to the audiologist's dog, a mostly black and very glossy-coated pit bull with warm brown eyes. She wants to check me out before I approach the front desk.

"Hello there," I say.

"She's a bit shy."

"What's her name?" I ask.

"Raven," the man behind the counter says. "She's kind of new around here."

"What's *your* name?" I ask.

"I'm Ray," he says. He has bright blue eyes and a shaved (or maybe naturally bald) head. He wears a black mask, and I wear mine. COVID-time, where only our eyes do the smiling. No one else is in the office. I fill out my paperwork while sitting on a cheerfully patterned couch.

When Ray comes around to my side of the counter, he gestures for me to stay in my seat. He wants to take a quick look at both of my ears with a handheld scope of the sort I've seen since childhood.

"You've got very small ears," he says right away. "I can see how they might tend to get a bit blocked. Very small."

I'm not sure what to make of this assessment. "Small ears meaning small canals?" I ask.

"Yes, and the outer ears too," he says, nodding.

It occurs to me that I've been told this before, about my small ears, though I can't precisely recall.

"Your ears look fine," Ray says, putting down the scope.

"You mean they don't need to be cleaned?"

"They really don't," he says. "I can see a bit of something on the right eardrum, nothing that would rinse out, not wax or anything. Hard to explain. But your ears are clean."

"So . . . there's nothing that's like an obvious blockage of any kind?"

"No."

I flash back to long-ago episodes in various doctors' offices where bits of wax had been noisily flushed with a giant syringe into an enamel pan. It's increasingly clear that nothing like that is going to happen today.

"I think you might want to get tested," Ray says. "If you're ready."

I sense he's deliberately trying not to pressure me. He must have plenty of experience with people who are terrified of finding out exactly how much hearing they have lost or are in the process of losing. I appreciate his patience, his understanding.

I say I'm ready. I want to know.

Following Ray and Raven to the back office, I see the dog's cushion in a place of honor, although she continues to sniff at me some more before she is prepared to relax in my presence. When I notice her prominent set of nipples, I ask if she recently had a litter. Ray shoots me a meaningful look. "When she got to the shelter, she had two puppies with her. So yeah, she's been through a lot."

"She seems very sweet," I say.

"Oh, she won my heart," Ray says.

Opening the door of the soundproof booth, Ray beckons to me. "Has it been a while since you've seen one of these?"

I mention that it has been a very long time since I've had my hearing tested, or my vision for that matter. I have always had excellent vision, twenty-twenty. I wear magnifiers for reading only, 1.5 or maybe 1.75. I'm proud of this, for some reason, though it really has nothing to do with any particular talent or skill. I'm just lucky. Good genes?

"It has definitely been a while," I say, taking a seat. "Finger up, to say *I heard that*?" My body is suddenly remembering how this procedure goes.

Ray hands me a black cord with a microphone-shaped object on the end. "We've improved things slightly," he says. "You push the button."

After Ray sorts through some items and finds earphones that are apparently small enough for my small ears, I remove my mask, since I'll be alone in the booth.

"Pure tones, right?" I ask him.

"Yes," he says. In my minimal research I've learned that this is the most common method for testing someone's hearing range.

"Being able to hear specific sounds in the presence of background noise is a more complicated kind of measurement," Ray says. He closes the door.

Both Raven and Ray seat themselves outside the booth while I attempt to calm myself. Through the glass window, I watch Ray put on a pair of headphones and start fiddling with some knobs and dials. I take a few deep breaths.

"Left ear first," Ray says, his voice finding me inside the booth.

The tones arrive in no particular sequence that I can determine, which I'm sure is the point. I press the button with certainty most of the time, wincing when I wonder if I am only imagining a sound. "Is this volume okay for you in general?" Ray asks. I nod.

He explains he's going to add some soft white noise in my left ear while he tests the right one—the one I am having trouble with. The tones start up again, and I feel my breath begin to tighten, recognizing a kind of performance anxiety, though it isn't Ray as audience I am worried about. It's my own ability to be an audience for myself. *Audition.*

I close my eyes. I'm wincing more, and I know it. I can feel the strain in my entire body, as if I'm leaning toward the sounds, as if even my eyes under their lids are trying to turn to the right, help my ear do its work.

My eyes are still closed when the booth door opens. Ray reaches over to remove the earphones and then invites me to come back into the office. Raven is curled up asleep in her bed.

"Okay," he says gently. "You've definitely got some loss. You weren't imagining it."

I sit down on a rolling stool while he prints the digital readout. I confess that because the machine on his desk looks like something from the 1960s, my mind begins crafting a set of ideas about the test's unreliability. *Maybe the results can't be trusted. Maybe I should have pushed the button even when I wasn't sure. Maybe I should go back into the booth for another round of tones. Maybe this entire visit was a bad idea.*

"Do you want to see?" Ray asks.

He holds out a single piece of paper with a set of graphs. *Audiogram.* There are his notated lines and dots representing my left ear's range, and then the ones representing my right.

"This is normal loss as part of aging," he says, pointing to the information for my left ear. "But in your right ear, you've got some sort of nerve damage." There is a fairly precipitous drop in the trajectory of my right ear levels; as the frequencies go up, my line goes down.

"But it happened all at once," I say. "Like overnight."

"It could be worse," Ray says. "Sometimes people lose even more, just out of nowhere."

I gasp, feeling increasingly desperate. "It's gotten better since that first morning, though."

Ray slightly tips his head to one side, his blue eyes fixing kindly on mine. "You've adjusted to it," he says. "You've compensated with your other ear."

My hands are clenching and unclenching themselves. I stare at my feet, fix my thoughts on the supportive floor beneath me. *Is it true? I've already adapted?* "What does nerve damage mean?" I ask.

"That means it's not coming back," he says.

What I read online, a day later, tells me a few more details that Ray didn't explain, but hinted at: "Damaged synapses are like unplugged connections, which especially affect our 'ability to make sense of complex sounds, and especially the ability to understand speech against a background of noise . . .' You can lose about eighty percent of the synaptic connections before that loss shows up on an audiogram."[3]

And also, this: "the ears you're born with are the only ears you get: a newborn's inner ears are fully developed and are the same size as an adult's, and, unlike taste buds and olfactory receptors, which the body constantly replenishes, the most fragile elements don't regenerate."[4]

. . .

Sometimes I say that I don't teach writing, I teach listening. My father, who observed a few of the workshops I lead, once commented that he was "impressed" by how well I listen to my students as they read their work aloud. Middle child in a clamorous family, I may have been born to be a good listener, though I also actively cultivate the practice. My laboratory takes place with small groups who come monthly to freewrite at my home. In these sessions, participants write in response to prompts, encouraged to write faster than they can think.

As for sharing the brand-new words they have just conjured, my instructions to the group insist on giving wholehearted regard to the person reading. *Not* to critique or even make suggestions but to tell the writer what they *love* about what they heard. To notice when, as listeners, they feel touched or awakened in some way; to give the most memorable words and images back to the writer; to let the writer know what *succeeded.* Attending so closely to everyone else also guarantees that there's no room left for worrying about your own turn to read. And then a kind of flow happens—because when the writers know their words are going to be affirmatively welcomed rather than judged or even "constructively" remodeled, they feel safe enough to explore whatever has been waiting to emerge.

. . .

In the new year of 2022, I'm in Tepoztlán, in the state of Morelos in Central Mexico, preparing to lead a three-weeks-long writing workshop as part of a conference called "Under the Volcano." Thanks to the persistence of the global pandemic, this residency is being offered in a hybrid format, the generic language we have come to agree upon for a mixture of in-person and virtual participation. I have six students in total: four are here with me in person, and two are connecting from their individual homes in Brooklyn and Houston.

The setting is idyllic, complicated, magical, and challenging.

The pueblo fondly referred to as Tepoz nestles so close to the mountains it sometimes feels as if you can reach out from your window and touch the craggy edges. A pyramid (from the fifteenth or sixteenth century) perches on a cliff above the town, inspiring weekend visitors to hike up an extremely steep rock-strewn trail. It's a pilgrimage for some, an act of bravado for others, who brace for the ascent with shots of mezcal sold by vendors on the road leading to the trailhead. The scene strikes me as a blend of New Age appropriation and macho partying. *Complicada*.

The first time I visited Tepoz, in 2019, I hiked up to the pyramid one early morning with a friend, and even though I considered myself to have been in great physical condition, the nearly vertical climb was exhausting. I kept telling my friend Cindy that it seemed as though the mountains were pushing back at me.

Alison, who lived for a few years in Tepoztlán while raising her son, is back in town and tells me that people here believe that whatever mood or sensibility you bring with you—whether it's emotional or physical or psychological—the mountains amplify it. This is literally true in terms of pure sound: almost every night at dusk, there are fireworks and explosions resounding off the rock walls, the noises stereophonic and echoing. A chorus of barking dogs and music blasting from church loudspeakers and wedding mariachi bands creates a cacophony that sometimes lasts until just before dawn, when the roosters take over. In other words, it's a loud place. Gorgeous. And loud.

The rented property where my workshop meets can best be described as an indoor-outdoor compound of open-sided structures made from concrete and stone. All of the participants, as well as the tech support team, have been vaccinated, and we all tested negative on arrival. The point is for us to be able to gather, maskless, in wide, doorless, airy rooms. Unfortunately, this also means that sound travels freely from space to space; I can hear the workshop "next door" to ours, and I can hear the chatter from the nearby kitchen as well as occasional laughter from people sitting beside the small pool in the central courtyard.

Because my teaching is based on deep listening, I'm straining to concentrate on my students: the four women sitting on a pair of couches situated opposite me, and the two appearing almost life-sized on the screen. My right ear is still not working well. The problem seems worse when I fixate on it, but I can't quite ignore it either, can't pretend this difficulty isn't happening. I suffer rising anxiety about how much extra effort it takes me to hear the words of my students; I keep leaning closer. And I have to ask, sometimes, for people to repeat themselves.

At the end of the first day, when Tim, an Irish writer who is leading one of the other workshops, mentions something about one of his students going to visit a local shaman, I feel an irresistible curiosity take hold.

Thirteen years earlier, in January 2009, I was traveling in Mexico not long after being diagnosed with breast cancer. The sound healer I visited in the city of Querétaro said he was going to play me an electronic tone that was meant for me alone, a frequency that would somehow align with the core vibration of my body, or perhaps my spirit. When he left the room (along with my dear friend Ana and her mother, who had brought me to meet him, and who served as my translators), I was left lying on the table where he had already placed various magnets near my feet and head, arranged along the outlines of my arms and legs.

As soon as it began, the tone felt so perfectly *right* that I burst into tears, and I continued to cry for the entire two minutes in which the vibration was sustained. I still don't know what the note was, but I do know it centered me completely.

For my appointment in January 2022 with Jaime, the shaman, I take a taxi several miles away from Tepoz. We creep along the cobblestone streets, over the unavoidable *topes* (Mexican speed bumps) forcing vehicles to slow down. Our route is leading in the direction of Amatlán, the village in which the Aztec god Quetzalcóatl, the Feathered Serpent, is reported to have been born. Eventually the taxi turns off on a side road that bears a sign for SITIO SAGRADO, which turns out to be a fancy walled-off hotel and spa (later, on the way out, I see an additional sign calling it a "Celebrity Spa"). Jaime's property is just beyond that so-called Sacred Site. He greets me at the gate in his all-white outfit, wearing Huichol beads and sandals. His dark hair and skin are in stark contrast to his gleaming teeth and sparkling eyes.

He says, "*Bienvenida, Elizabeth,*" and asks the taxi driver to return "*en una hora y media.*" Inviting me to take a seat in the shade of a tree, he also points out his *casa con baño* located down a small pathway. "*Estoy con otra persona,*" he explains, gesturing back to where I see he had been sitting with another client, a woman whose back is turned. "*Dame un ratito,*" (Give me a little while).

I relax immediately in the dappled light, soothed by the sweetness of a warm breeze. After a few minutes, though, realizing that maybe I should use the bathroom while I have some time, I wander over to the house and knock on the door. When there is no answer, I let myself inside and am startled by a young man calling out from another room. I stammer and profusely apologize, embarrassed to have intruded, but he is kind enough to simply point me up a flight of stairs. "*Adelante.*" (I find out a few days later that he is Jaime's son.) There's an elegant bathroom with tub perfectly situated to face the mountains, and a mood of tranquility everywhere.

Back in my shady seat I notice a brown dog half-asleep in the dust, barely opening one eye to acknowledge my presence. Was he there all along? I vaguely perceive the edges of the firepit in the open space of the ceremonial hut and the murmured voices of Jaime and the woman he is working with. I wait and doze, listening to the wind.

Eventually, I see Jaime and the woman walking out together to the gate. She gives me a small wave; he comes back saying, "*Elizabeth, vamos. Bienvenida, otra vez.*" I explain that Spanish is my rather inadequate second language, and as he instructs me to take a chair by the firepit, he adds another log and says we can begin with English, that he will do his best and see how it goes.

"First, we will sit together with Grandfather Fire."

Aware of the nervous effort of holding myself up, I lean back into my seat, as Jaime appears halfway reclined in his. Allowing the flames and smoke to speak to me, absorbing the glow of the embers, I discover how terrified I am to feel my identity as a listener, the heart of everything I am and everything I do, possibly being lost. It's not the smoke that brings tears to my eyes, but my overwhelming dread. When I take a tissue from the box beside me, Jaime says, "Okay, so tell me why you are here."

I try explaining in just a few sentences about this partial hearing loss on my right side, about how it threatens my work as a writer and teacher, but also my sense of self, my core. We nod together at how obvious it all seems, the collision between a literal symptom and what it seems to represent.

"I grew up with so much fear," I add. "Holocaust-survivor parents who taught me that there are dangers all around, lurking. I know there is vigilance that's useful, protective, and then hypervigilance that is paralyzing." Jaime continues to nod, watching me closely. "I also know that sometimes it's hard to tell the difference."

I want to remember his words, even though at the end of our session he tells me not to worry about the words so much as the experience *inside* the words, inside my body (even those are not his words, exactly). But I want to remember the first things he said about the Huichol belief that "in the beginning is the Sound, the origin of everything is Sound, and this is represented by the Blue Deer."

"What did you say?" I ask.

"The Blue Deer," he repeats. And then he points to Grandfather Fire, saying, "This is the heart," and much of what he says next is about listening with the heart, listening to the heart, trusting the heart not the mind, knowing that fear is in the mind, understanding the difference between the fear that is protecting us and the fear that is a distraction.

The heart, the heart. Fire. It's everything. It's connection. Heart connects us to fire, connects us to everything, to each other.

He talks a lot about connection, says he can tell I am *connected*. He assures me that I know what I need to know, and that I am a flower with petals still opening. I still have much to live, much to give, much to offer to the world.

He talks about saying *Yes* to what I've been given to do and to be, and that this may not be the same as what I think I need or want. He compares it to his own work as a shaman (he uses the Huichol word, *marakame*). About how hard it is and yet we have no choice, this is who we are and what we are given.

"Huichol are very practical people," Jaime says, and he talks about humans in our distinction from animals and the rest of creation, how we are the only ones with awareness to ask questions and to have uncertainty about our purpose. "The tree doesn't wonder where to throw its shade, the bird doesn't ask if it is supposed to be a bird." Our blessing and our curse (my words) is to have been given the questioning mind. "Otherwise, we wouldn't be here at all, we wouldn't make it," he says.

"Do you listen to the wind?"

I nod.

He asks how long I have been writing books.

"Thirty years," I say.

He says, "You have already spread many seeds then. And you still have more." He says he has been doing this work for thirty years also. Explains that he came to Mexico from Colombia for what he expected would be five days, and he has been here for forty-four years.

He tries to explain the difference between feeling lost and feeling like *I don't know where I am*. That is when I must listen to my heart, he says. I must trust in the sound of the wind, the messages of connection, the voice of Grandfather Fire.

At the end comes the *limpia*, with Jaime using feathers to dust and sweep across my face, shoulders, back, arms, head, ears, legs, feet. He also gives me a crystal to hold in my right hand. I'm seated closer to the fire now, barefoot and without any metal (jewelry). I feel something being sucked out of me (literally: Jaime presses a straw against different places on my skin) as if a poison is being removed, something to set me free. Maybe what's being extracted is my fear, and maybe it is my inherited fear, and maybe it is something like grief, I'm not sure.

A few days later, at dinner, I find out that Alison is best friends with Jaime's wife, Lourdes, and she went to Jaime for a series of sessions some years ago. In fact, she knows their whole story, how he came north from his birthplace in Colombia and joined an eco-community here in Tepoz, was walking on the road to Amatlán and was hit by a speeding truck. Terribly injured, he was almost killed by the truck driver who was about to run over him a second time, Alison says, "to keep from being sued." But at that moment a van full of blind children appeared, and they became "witnesses because they *heard* everything," especially the dramatic shouting of their own driver who saw what was happening, who saw the truck aiming at the broken body of the man on the side of the road.

"Jaime remained in the hospital for months, with so many broken bones and scars," Alison continues, and then after that, he and Lourdes met at a party. "She is a doctor, a homeopath, an acupuncturist, very pale and contained and serious, while he is a jaguar in human form, dark and full of coiled energy." Yet they came together instantly. "Everyone said it would last a few weeks, they were so different, so opposite," Alison laughs. "And here they are, they have a son, they are still together!"

Adding to what he told me, Alison explains that Jaime was chosen by the Huichol to be a *marakame*; he didn't want this but they said he had no choice, they had plans for him, and for twelve years he was trained and initiated, went on pilgrimages and journeys all over. "He is the real thing," Alison concludes, "and so is Lourdes. There are people in Tepoztlán who invent themselves and claim all sorts of things, but he and Lourdes are the real thing."

Later that night, my final night in Tepoz, while asleep in my bed, I am suddenly infused with the distinct aroma of smoke from sage or an echo of Jaime's sacred fire, one intake of breath through my nose, unmistakable, just that once, followed by cool night air. The sensation is strong enough to wake me up, as if I have just received a blessing, a message.

. . .

December 2022. My second hearing test, almost exactly one year after the first one, is scheduled again at the Alameda audiology office. I remember the sweet dog named Raven; does she remember me? Does Ray remember me? We are both still wearing COVID masks.

A lot has happened, in a way. I've seen an ear, nose, and throat specialist who, after confirming that there was nothing wrong with my outer ears, ordered an MRI as well as another updated measurement of my hearing. The MRI results showed the presence of a benign tumor in my inner ear, a *vestibular schwannoma.*

A new audiogram will help show in more quantitative terms how the tumor may be further impacting my auditory nerve, and will give an indication of how fast the tumor is growing.

"I'm sorry," Ray says, when I share the news about the tumor, adding, "I mean, I'm sure it will be okay." He tells me he is sorry he didn't tell me a year ago to get an MRI, but we look at each other with expressions of acceptance; we both shrug. I interpret his wordless communications to mean the same as my own: Would it have been any different for me to get the same news a year ago?

This time, though, I'm even more scared of the results of the hearing test. Because they might reveal the potentially rapid rate of growth, leading my thoughts to worst-case scenarios like possible radiation, possible surgery, possible outcomes like total loss of hearing on my right side.

After the second test, after the déjà vu of sitting in the booth, pushing the button, and straining to decipher what I think I am hearing, Ray prints out the information and says the good news is that I am retaining my clarity. He had recited a bunch of words, one at a time, and I was to repeat them, one after another, *sled*, *greyhound*, *carve*. I was listening to myself say the words very deliberately, every vowel and consonant, the hard ones and the soft ones. It's only afterward that I think about the way I've always prided myself on my enunciation, on not having *an accent*. I think about those times I tried correcting the pronunciation of my father, teaching him to say *three three three* not *sree sree sree*, and how he stuck the tip of his tongue between his teeth, trying to imitate me or maybe make fun of me, and how when I tried to imitate his pronunciation of the number seven in Swedish, he tried to show me how to shape my mouth and it just wouldn't conform.

Eye *tests*. Hearing *tests*. The word itself creating an association with performance, with measuring up, with being graded according to the standards of perfection, someone else's idea about what's normal, what's best.

And what is my own new normal going to be? The good news is that the tumor is growing very slowly. For now. The good news is that it is benign. "The bad news," says the second-opinion ENT, when we meet via Zoom—while I'm back in Tepoztlán, teaching again in 2023, looking up at the echoing mountains so close and so far away—"the bad news is its location, the worst possible location." The doctor says this with a gentle smile, and shows me an MRI of a brain, not mine, for some reason, but someone else's MRI, and then he shows me an anatomical drawing of where the tumor is, how my *vestibular schwannoma* is located in the narrowest part of the canal. How it doesn't have anywhere to go except to squeeze outward, into the cistern.

And even though I'm listening to the doctor I'm also remembering the haunting sounds of water dripping inside the cistern I visited in Istanbul decades earlier, the place known as Justinian's Underground Cistern. He was the renowned sixth-century Ottoman emperor who had commanded it to be built. I can vividly remember how, walking along ledges inside the cistern, I listened to the water dripping in counterpoint with the piped-in music playing there, something elegant and beautiful from Verdi. I'm thinking about Pauline Oliveros, who invented the term "deep listening" after recording the echoing reverberations in a cistern nearly twenty feet below the surface of the earth. And then I think about the *cisternas* all over Mexico, including here in Tepoz, where people store their water, because we're in the desert, in the mountains, and sometimes it doesn't rain for months, and other times it rains and rains and rains.

Do you listen to the wind?

It doesn't rain but it pours. Like good news, like bad news. Like the sound of wind suddenly threatening to blow the branches off the trees, the way the wind rattles the windows and doors, the way it sounds like the wind is trying to get inside the house, inside my head. The way in 1983 in Germany at the top of the mountain called the Ettersburg, the wind carried the unmistakable sound that my father and I both heard together as we were ending our first visit to the deserted yet haunted grounds of Buchenwald. I will never forget the voice of my father saying to me, "Does it sound like people screaming?"

ALMOST-PERFECT RECORDINGS

Rewind.

When I was very young, we owned a reel-to-reel tape recorder. Around age ten, thanks to my father playing some of our old tapes, I got to hear the sound of my voice at the age of two, alongside the voice of my older sister.

"This summer I read thirty-seven books from the library," Monica announces. "I'm four years old."

"I'm four too," my high-pitched voice insists, but my sister cuts in, "No you're not. You're only two."

I don't know what happened to those reels of tape, those voices. They've been erased by time and by loss. But somewhere inside me I can still hear us.

Rewind. Record.

I'm twenty-two and visiting my parents and younger brother in a suburb of Toronto, where they've recently moved. My father gives me a miniature handheld tape recorder with a few blank cassettes. Once upon a time he worked in a research laboratory where computers filled entire rooms, and he has always loved collecting devices like this one, marveling over the improvements in technology.

I can't quite get used to the fact that my family lives in Canada, though this relocation will turn out to be temporary. My mother hasn't yet been diagnosed with bipolar disorder, and we are still tiptoeing around her unpredictable mood swings. Everything about the visit feels surreal to me: familiar furniture spread out in new arrangements; a kitchen in which I have to open every cabinet before finding what I need; a guest bedroom for me to sleep in. It's as if I'm a foreigner, an alien, my own island in the archipelago of my family.

My father says he wants to discuss something. "Maybe you can try out your new recorder," he suggests.

It's late summer, so we sit together under an awning on their new back patio; the scraping sound of aluminum-framed lawn chairs on a concrete slab brings back a sense memory from decades earlier.

"What do you want to talk about?" I ask him.

Click.

And although that tape too has long since vanished, I saved the conversation in my mind. I remember the shape and meaning of what he said. I remember what I heard.

My father wants to explain to me his theory of parenting; he wants me to understand how he figured out what kind of father he tried to be. "You know, I basically had no parents," he begins. "My mother and father divorced when I was only seven, and not long after that I was placed in an orphanage with one of my brothers. We had people taking care of us, but not really *parents*. So, I had to figure everything out on my own."

Rewind.

I can visualize him in the orphanage because I've seen a handful of black-and-white photos from that time in my father's early life. Nazi Germany, a Jewish orphanage in Hamburg, the horrors of a concentration camp lying in wait just a few years into his future. But meanwhile, he's still a boy. At least one photograph shows him with a protective arm slung over the shoulders of his younger brother, Joseph (then called Wolfgang). Figuring everything out on his own.

"I used to daydream a lot about what it would be like to have parents," he continues. "Especially to have a father guiding me and helping me. Someone with enough wisdom and experience to tell me what to do."

This is where my memory fades. Did I ask him questions? Did I interrupt him before he finished? I am superimposing my current awareness onto the scene again, wondering how much of what I was listening to made sense to me at the time, what it might take to add my own layers of understanding. Because eventually, I recognized that my father tried to become the father he wished he *himself* had grown up with. In his longing for an all-knowing parent to provide him with protection and security during those tumultuous years, he promised himself he would give to his future children what he never had.

Yet there I was, with a different set of needs, someone who *enjoyed* figuring things out on my own, a daughter who felt oppressed and even silenced by the all-knowing father. The same rules and structures he imagined would have made his own childhood safe were the ones that made me feel unsafe. Unseen. Unheard.

Fast Forward.

The day after I fly in from the West Coast in early April 2022, my father is in pain from what is likely a severe bedsore, and he's desperate to see the doctor. I'm desperately apologizing because the appointment is too late in the day, and I just can't get him there on my own. The aide has left early, and I'm alone, and I'm not strong enough. It takes only one person to wheel him into the doctor's office, but it takes more than one person to get him into the car from his wheelchair, more than one person to get him out of the car and back into the wheelchair again.

"I can help you!" he says. "I can do it myself."

My father still stubbornly thinks this way. It's likely that he's not confused so much as in denial, because this kind of denial is the story of his life. Dissociation, some might call it. Separating the head from the body that has carried the head around all these decades. I'm convinced it's my father's survival strategy—originating from the year he spent as a teenager in a concentration camp. And probably from before that, even. From the years of his childhood abandonment, from his near starvation. From bombs. War.

I have to gently remind my ninety-three-year-old father who can't stand or walk on his own that he's mistaken. He's not strong enough.

"I'm so sorry," I say again. "We have to wait until tomorrow when we have some more help. Please forgive me."

"Of course, I forgive you," he says, sighing. "I always forgive you. There's nothing you can ever do that I won't be able to forgive you for."

Rewind.

"You're probably right," my father admitted to me once, when we were revisiting the angriest years of our past, when I reminded him that if we as a family had changed our synagogue membership from the Orthodox to the Conservative, everything might have been different. There, among other more egalitarian practices, the women and men sit together.

Even though my mother had tried many times to explain to my father that he was looking in the wrong place for evidence of my Jewishness, that it was on bold and poetic display in my writing most of all, he had remained dissatisfied, disappointed. She herself always insisted that she was "Jewish on the inside," when he accused her of setting a bad example for me, because of the way she flouted the rules of *kashrut* by eating shrimp cocktail at restaurants, by going shopping on Shabbat mornings instead of attending synagogue with us. And then, to make matters worse, around the age of fourteen I insisted on joining her for these excursions, refusing to trudge the mile of distance with him and my siblings, shouting that I would no longer agree to my second-class citizenship during services at *shul*.

In the years past my mother's death, though, my father began to see and hear what she had already seen and heard, that my own version of *Jewish-on-the-inside* was being expressed in my books—all of which had been published after she died. He discovered it was possible to be proud of me. Sitting together in his den, surrounded by piles and piles of magazines and books, his obsessive hoarding of *papers papers papers*, we had stopped yelling at each other.

Which is why, when he said those words, "You're probably right," I felt something soften, like a chord resolving from dissonance. After so many inadvertent (and perhaps even deliberate) wounds we inflicted over decades, we were learning how to have quiet conversations, listening to each other with mutual attentiveness and forgiveness.

"I'm sorry," my father said.

"I'm sorry too," I said.

Fast Forward.

"Welcome, welcome, welcome," my father says. His face is lit with joy. "I'm so glad you made it," he says into my ear when I bend down to kiss his cheek.

He's in his chair at the kitchen table, the executive leather chair on wheels that used to be behind his executive desk at the office. He's eating spoonfuls of soup. Sipping.

I'm dragging one large suitcase, one carry-on suitcase, and the ghost of my dog Lulu, who had flown across the country with me in June 2020, and who died here without any warning two months later. Now it's April 2022: my father and I have both made it through two years of COVID.

"I'm glad too," I say.

"I feel so confident now that you're here," he says. For some reason I don't ask what he means; it's as if I already know. As if we both know.

An hour later, he's tucked into his hospital bed in the living room, the details gone over, the *papers* in their place. (Pockets of his pajamas are stuffed with tissues he has torn into small pieces that he calls his papers. The behavior was maddening until I realized that this isn't just about a constant need to wipe his dripping nose. This is some unconscious revisiting of a time when having—or not having—the right documents, the right *papers*, meant life or death. We don't discuss this.)

"Everything perfect?" I say.

"Oh, Elizabeth," he says. "Nothing's perfect."

I laugh. "Is it almost perfect?"

"Almost perfect," he says.

We laugh together. His aide, Lenore, laughs too.

I kiss him on both cheeks. "Good night," I say. "I love you."

"I love you too," he says. "I'm so glad you're here."

I sit quietly at his bedside for a while. Lenore has disappeared into the kitchen to wash the dishes before returning to her home for the night. My father's eyes are closed, and it's dark in the room. Dark except for one night-light in the corner.

"I've been thinking," my father says, his eyes still closed. "About what it would be like if I died while you were here."

I take a breath, carefully inhaling and then exhaling. "It would be okay," I say. "We would get to say goodbye to each other."

Pause.

My father is the person I fought hardest against throughout my childhood and adolescence, at times even during my adulthood. The person I yelled at the most and the one who yelled the most at me. Yet, for the past decade, maybe more, we have been saying "I love you" to each other as if it's the easiest phrase in the world. I believe I taught him how to say it, never hearing him say these words to my mother in their almost fifty years of marriage. Of all the questions I asked, I didn't ever wonder aloud whether he and my mother said *I love you* to one another. Maybe that was one of the Swedish secrets they kept from us. *Jag älskar dig.* Maybe they kept this secret from each other.

My nephew Ezra—the one who has made a PowerPoint presentation based on my father's stories of survival during the Holocaust—calls my father every single evening. He always ends the conversation by saying "I love you, Grandpa," or sometimes "I love you very much."

"I love you too, Ezra," my father says, every time. I wonder if Ezra knows that this is the same person who used to terrify all of my friends with his anger, the same person who used to terrify me sometimes. Maybe my sister, Ezra's mother, has told him about what our father's rage used to sound like when we were young. Maybe she doesn't remember him the way I do.

Fast Forward.

During an early morning walk in my father's neighborhood, dogless and solo now, I notice an exuberant labradoodle. She is pulling hard on her owner's leash to greet me, and I ask if it's all right to say hello.

"Of course," the woman laughs, "since Rosie obviously loves you already."

When I stroke her back, the joyous dog's pale-peach curls feel like cashmere.

"What's your name?" I ask, after telling her mine.

"Helena," she replies, and asks if I live nearby. I point to my father's house and add a few words about his ninety-third birthday a few days earlier, about his dire condition.

She nods. "I've noticed what must be caregivers coming and going for quite a while."

The empathic tone of her voice enters me right away, and I feel tears coming. "Yes," I manage. "He's been living here alone, and I live in California, and my siblings are closer but in other states too, and he has so many health challenges . . ."

Helena's eyes fill with tears. "I understand," she says. "My mother isn't as old as your father, but still, she is far away."

. . .

Did I ask where she was from, this beautiful woman with such a kind heart and some kind of Eastern European lilt to her voice? I must have been trying to guess, the way I always did when I heard a certain sound, an almost-recognizable melody. "Prague," she must have told me, either in that conversation or the next one, when we met again on the street just a few days later.

Except by the time of that second encounter with Rosie and Helena, my father was gone. I had entered the next part of my life, the rest of my life, the part where both of my parents were dead, and I would speak of both in the past tense—not past tense for my mother (who died more than twenty years earlier) and present tense for my father—not this distinction ever again.

I told Helena, and we both cried, and she asked me what she could do, could she make me a cup of tea or give me a hug or both, what did I need, she was right there, a neighbor, a daughter, a fellow lover of dogs.

In Helena's elegant kitchen, the one she designed after she and her husband had purchased the house on my father's street, the one with "good mid-century bones," I marvel at the beauty of the open spaces, the way afternoon light pours in from big windows facing a swath of green yard at the back. They had moved in during the pandemic, and I must not have noticed all the construction going on down the street from my father's house; we were all so insulated from each other, so self-absorbed. Helena mentions she is still commuting up to Lake Placid where they moved from. "I love the mountains," she says with a wistful smile.

She tells me about how she keeps trying to adapt to this flatter landscape, this one that is more like a prairie than where she grew up and also where she recently lived. Niskayuna, the name of my town, translates from the Iroquois language as "land of waving grass."

"I know there is plenty of grass around here, plenty for a goat to enjoy," she shrugs. "But I'm a mountain goat."

When Helena says she is sorry she never got to meet my father, I tell her that my mother would have especially loved her, my mother who adored dogs and cherished her many friends from Eastern Europe. I tell her about my friend Zdena Berger, who also came from Prague, who wrote a book about surviving four years in concentration camps. I explain about the one time I visited Prague in 1995 with my parents and siblings after we went together to visit Buchenwald for the fiftieth anniversary of its liberation. I tell her that my mother, who spoke many languages with such fluency, found it almost impossible to understand Czech. And I tell her that although my parents and my sister flew back home to celebrate Passover, my brother and I stayed on for a few extra days because we were infatuated with Prague's Bohemian charm.

"We rode the underground all over the city," I tell Helena, and then I can't help smiling because I'm remembering how my brother and I memorized what sounded to us like a seductively melodious announcement. "We would chant it back and forth to each other while riding that long, long escalator up and down, and all of the other passengers must have thought, 'What's wrong with those two! Why are they telling each other that the doors are closing, over and over again!'"

Then, sitting in Helena's light-filled kitchen across the street from my parents' now-empty house, I repeat the words which I've never forgotten, as Helena laughs with me: *dveře se zavírají*. The doors are closing.

Helena pours tea into a gorgeous teapot, setting out cups and saucers, pink porcelain with white-and-green floral decorations, delicately gilded edges. "Yes, this set comes from the Czech Republic," she says sadly. When I tell her my story about being told my parents had accents, she looks even sadder. "The same thing happened to me," she says, "except I was the mother with the accent."

Her children's friends were visiting, and her daughter's friend from school said, "I love the way your mother talks," and her daughter said, "What do you mean? How does my mother talk?" and the friend explained about the accent, and her daughter said, "What accent?" and Helena says she realized for the first time that, to her children, she sounded "normal." That is the word she uses when she tells me this story, *normal*, and her eyes brim with tears as she explains that people in the United States always ask where she is from (as I probably did). And to her children, at least up until that moment, she didn't have to explain anything.

Record.

The world tilts again, or maybe rights itself, or simply grows more full. I am not a mother, but I will always be a daughter, even with both parents gone. I will never again hear their voices except in recordings, in flashbacks, in my dreams. Soon, I will listen over and over to the voicemail messages I saved from my father, sorrowing anew about not having any such treasures from my mother. I will notice, in watching both of their Shoah Foundation testimonies, one at a time, two hours each, how my mother's accent returns me to the sound of her voice I thought I'd forgotten. But it seems to me that maybe my father's accent continued to change over time, over a stretch of twenty-five more years. That video captured him as the father I no longer precisely remember, with the voice of my younger father.

The sound of Zdena's voice and the sound of the Prague subway announcer and the sound of Helena's voice and the story of her children listening to her are intermingled now with the story of my father's death. I have never until today wondered what it was like for my mother or father to be told they had accents, or if I ever told them what my friend said to me. I realize that there is no such thing as the sound of "normal" except for what any of us decides it means. I hope that from now on, anytime I ask someone where they are from it won't sound like a criticism or even an insensitive remark but a genuine desire to connect, to say *I want to meet you*—the world, this life of everything everywhere—with welcoming ears. *I hear you and I want to know about you and I love the way your words are shaped in the language of my own home but also blended with the sounds of where your own home began.*

. . .

Shemá! Listen! That pleading and demanding first word, followed by a sequence of devotional promises to be recited at least twice each day, upon waking and retiring—and also when going out of the house and upon returning home. Inscribed on your forehead and affixed to the doorposts. Some say that the prayer is meant to be exhaled as the last words spoken by a Jewish person at the end of life.

Pause. Record.

On a yellow pad, I wrote down this conversation with my father, adding the date at the top of the page: April 13, 2022. I'd been sitting by his bed for hours, mostly in silence, holding his hand. Since his eyes were closed, I thought he was asleep, but then he asked me a question without opening his eyes.

"You know what you say when you want to forget something?"

"No," I said. "What?"

He opened his bleary eyes for a moment to try focusing on mine. "I forgot," he said.

I smiled. "Okay," I said, and he closed his eyes again.

More silence passed. The morning light pushed through the gauzy curtains in the room, and I was about to get up to make a cup of tea when my father spoke again.

"*Sel-lah*," he said, slowly emphasizing the two syllables.

"That's the word?"

He nodded, then spelled it out for me, "*Selah*."

"Okay," I said.

"Thank you," he said.

Later, when I looked up the word's meaning, I found that although it's used seventy-four times in the Hebrew Bible, its etymology and precise meaning appear to be unknown.[1] Here are some of its uses and interpretations:

To pause and reflect;
To lift up;
A musical interlude;
To stop and listen.

Rewind. Record.

Some years before my father died, I don't remember exactly when, I'd gotten one of those terrifying phone calls from my sister or my brother, letting me know that our father was in the ICU again, at Albany Medical Center. I flew across the country, terrified of arriving too late. My mother had died while I was on my way, in the air, in the clouds. But with my father, whatever year this was, I got to the ICU, he was still alive. In fact, he was already beginning to recover.

"I was so scared," I admitted to him, sitting next to the bed in the curtained space of the ICU, watching all of the monitors beep and blink. "It's frightening to be so far away . . . and because of what happened with Mom."

"I'll wait for you," my father said, turning his head to look straight into my eyes. "I promise. I'll always wait for you."

I knew it was a promise he could make with total conviction and yet also a promise he might not be able to keep. Because who can know? Who can promise such a thing?

My father once told me that he had tried to convince my mother that she shouldn't be so pessimistic about her breast cancer diagnosis at the age of seventy. "Your mother lived to the age of eighty-six," he reminded her. "And your aunt lived to 104."

"Don't curse me," my mother fired back at him. "I don't want to be an old lady."

"She got her wish," I said to my father when he recounted the story. She died so suddenly that only my father and sister got to be with her, and none of us really had the chance to say goodbye. Her body abruptly shut down just five months after her seventieth birthday.

Play.

Although reading had always been one of my father's greatest life pleasures, his eyesight began dramatically failing in his late eighties, and reading became difficult. My sister or I occasionally read aloud to him from the absurd stacks of newspapers and magazines (for which he continued his subscriptions). But during my long stay with him in 2020, we decided to listen together to the audio version of my most recent book, *Survivor Café.* I'd recorded it two years earlier, at a sound studio in Albany, not far from the house in which we sat together.

While my father was propped on pillows in his beloved recliner, I perched on a chair beside him where I could keep track of the laptop controls and also watch him for signs of weariness.

"My eyes are closed but I'm still listening," he said a few times, when I pressed the stop button after thinking he had fallen asleep.

"Okay," I said. "But how will I know if you're awake or not?"

"I'll tell you," he said. And we laughed.

Because so much of *Survivor Café* features trips I made with my father to visit the site of the former concentration camp of Buchenwald, journeys that stretched across decades from 1983 to 2015, it was more than a little remarkable to sit with him now, this way. My sense memories felt eerie and entangled, as if we were taking these journeys side by side in repeating loops of time, as if I were both participating and narrating. Every once in a while, my father wiped at his eyes with some of his ubiquitous supply of tissues or released a deep and wordless sigh. I'd wanted so much for the book to be published while he was still alive. And now we were listening to the words together.

In addition to the audiobook, we listened to some of my father's favorite CDs, mostly opera and classical music, occasionally some relatively modern soloists like Edith Piaf and Frank Sinatra. Tipped all the way back in his recliner, he would visibly relax so much that I could see age lines disappearing from his face. Maybe he was recalling the times he and my mother heard these pieces together, in the early days of their marriage.

When I found some videos online of Leonard Bernstein conducting Beethoven's Fifth Symphony, we watched several performances together, my computer balanced on a stack of pillows on my father's lap.

"Isn't he amazing?" my father kept exclaiming.

I agreed. He really was. We must have played the one we liked best—with the Vienna Philharmonic—a dozen times, maybe more, marveling over the rapturous, serious, tender expressions passing across Maestro Bernstein's face. We took turns pointing out to each other the way he conducted with his entire body as well as his eyes, even his eyebrows. His hair.

"Absolutely amazing," my father said.

Play.

Beethoven composed his Fifth Symphony during a time when his deafness was notably increasing. He wrote to his brothers that this had "brought me to the verge of despair." Many writers, including novelist E. M. Forster in *Howards End*, have described the symphony as depicting a battle between the forces of darkness and light, represented by prolonged and uncertain tension between major and minor harmonies. Forster's narrator, at the very start of the fifth chapter of *Howards End*, refers to the symphony as "the most sublime noise that has ever penetrated into the ear of man," while his characters hear heroes and shipwrecks, goblins and elephants, panic and emptiness.[2] When the final movement "produces a bright, major-key melody," it is understood, at least by some, to signify "persevering in the face of hardship, with contrasting harmonies and melodies acting as metaphors for life and death."[3]

Resisting the temptation to end his life, Beethoven wrote: "Only Art it was that withheld me, ah it seemed impossible to leave the world until I had produced all that I felt called upon me to produce, and so I endured."[4]

Rewind.

Retirement was a word we never used with my father, even though his cumulative physical ailments by the age of ninety prevented him from going into the office any longer. Thanks to the service of Cathy, his executive secretary—and in many ways his closest friend—he persevered with "running the business" from home, emphatically refusing to give up on his plans for revolutionizing the diagnosis of heart disease by way of cutting-edge technology known as magnetocardiography. Having worked for more than seventy-five years of his life (especially if you were counting his slave labor in Buchenwald), my father didn't gravitate toward anything "merely" pleasurable.

Somehow, though, in addition to helping rediscover his love of music, I was shocked to notice that he had begun to enjoy watching birds that visited the railing on the back deck of his house. Until then, I had been convinced that the natural world was ignored by my father on his way to and from work, in a manner almost as extreme as the way Oscar Wilde defined nature as "a place where birds fly around uncooked."[5]

I started putting out breadcrumbs to attract more of a crowd; these offerings were eagerly consumed by squirrels as well as an assortment of blue jays, doves, chickadees, and cardinals. As a surprise gift, I picked up a used bird feeder from a local garage sale and set it up where it could be viewed from the kitchen table, the place where my father took most of his meals.

The appearances of the eye-catching cardinal seemed to delight him especially, and he caught me completely off guard by suggesting that we should give it a name. My father, after a full lifetime of seriousness, had turned belatedly playful. Cathy, who was visiting at the time, mentioned that some people believe cardinals embody the spirits of visiting relatives who had passed away. So, I asked my father who he thought the cardinal might be.

Without hesitation, he said, "I'd like it to be my mother."

I studied his pensive expression, and said, "You used to call her Mutti, right? Since it's a male, maybe you could call him something close to that . . ."

My father smiled. "How about Mooch?"

And just like that, the cardinal was granted not only a name but also a personality. We watched for Mooch so often that we soon realized Mooch had a mate, whom we called Mrs. Mooch, and eventually there were multiple Mooches stopping by (a family? a few best friends?). Lenore, my father's aide, said she could always tell when a cardinal was nearby because of the sharp chirping notes that she heard, like an announcement from the nearby treetops.

"Mooch is coming," she'd say.

Fast Forward.

The antibiotics prescribed for my father's infection are brutal. After taking them for a few days, he doesn't want to drink or eat anything; he has diarrhea, and we have to keep turning him as we change the bedding. He hasn't left his bed at all since our visit to the doctor. When he cries out in pain, I apologize over and over. *I'm so sorry I'm so sorry.*

He's not speaking anymore. My brother and sister and all of their children have driven here, flown here. We tell my father we are all gathering to celebrate Passover together.

It's April 16, 2022, morning. Morphine drops have been prescribed. The entire family stood around his bed the night before, singing for the start of Passover—which also happened to be the day of my brother's birthday.

We filled the house with food and rituals and prayer for that one full-moon night, connecting to the distant past and the near past and the uncertain future. Without opening his eyes, my father managed to sing a syllable or two along with us, in honor of the seder. In honor of the story of exodus and liberation. His breathing alternated between very shallow and very raspy, very labored.

For most of the night, he slept. We all tried to sleep. From every room in the house, we were all listening to the sounds of him trying to breathe.

And so that morning, while the house is momentarily still and I'm alone with my father, I hold his hand. I'm whispering.

"Thank you for everything," I say, leaning close to his pale face, my voice breaking. "Thank you for waiting for me. Thank you for keeping your promise and waiting for me. I love you. And I know you love me. I heard you. I heard you."

My father squeezes my hand. Once, twice, a third time.

Record.

I wanted him to know with certainty, as he was leaving, that I carried inside me all the things he tried to tell me during the six decades of my existence. Because the truth is that I had always been listening, even when I was so furious at feeling unheard that I pretended not to listen, even when I held my hands over my ears, even when the two of us shouted louder and louder to drown each other out. Maybe he had always been listening to me too, and for a very long time he didn't know how to show it.

What I believe now is that my father and I both found each other underneath words, in the silence beyond words. We had arrived, together, in a place of completion, in these final minutes of listening to each other on the earth. We had made it all the way around the circle, from before my birth to his death, and perhaps even after his death toward the eventuality of my own death. All the way into my own future silence where, if I'm lucky, someone will be listening to me breathe my final breaths too.

Selah.

Rewind.

Three weeks after my mother died, in November 2000, I flew to Mexico to spend a month in San Miguel de Allende, a place I'd been hearing about for years, a place that had become famous not only as a UNESCO World Heritage Site but also as an unofficial sanctuary for expat creative types from around the world.

My mother's death had come unexpectedly, and even though I decided to go ahead with my preplanned trip, I had no idea how I would feel once I got there. I told myself I could pack up and return home after a few days if everything seemed too hard and too lonely. If I simply felt too sad to stay. Still, it seemed possible that the high-desert landscape of Central Mexico would be exactly what I needed. Maybe being in a town (and a country) where nobody knew me would give me a space in which to grieve with abandon. Maybe uncertainty and loss and exile and possibility would become part of this next chapter of my life.

On the shuttle from the León airport to San Miguel, I met a group of Americans and Canadians who invited me to a birthday party being celebrated at a villa in the Centro the following evening. "A great way to meet people," they assured me, "especially some artists." That same day, I found an apartment I could rent for the month, featuring a small but very charming interior courtyard with a spiral staircase I learned was called a *caracol*, or snail shell—a word which seemed perfectly suited to the way I felt I was carrying my fragile home around on my back.

The timeworn cobblestone streets and colonial architecture suffused with enchanting light soothed my spirit, at least somewhat. The birthday party turned out to be for a local celebrity whose garden contained a number of birdcages holding parrots and macaws. What I remember most from that night is staring into the eyes of each bird, startled and yet reassured to feel as though they were holding my gaze with tremendous intimacy. In a country where the dead, despite being invisible, are considered to be fully present alongside us, I sensed my mother's spirit all around me. Especially in the eyes of these colorful, intelligent birds.

On my second day walking around town, every arch and doorway and mosaic seemed to feature images of the Virgin Mother. I wandered into a gift shop whose name I recognized as part of a global organization called Save the Children; it was filled with handicrafts, and a woven-cotton shawl the yellow of an egg yolk caught my eye. Resurrecting my rusty Spanish, I asked the woman sitting at the cash register where the beautiful shawl had been made, and then immediately she asked me where I had learned how to speak Spanish so well.

At first, I explained about studying in high school years earlier, followed by a year as an exchange student in the Philippines. When I added that my mother spoke seven languages, I burst into tears. "*Mi madre murió*," I managed, "*hace un mes*." My mother died one month ago.

The woman beckoned to me to come closer, then pulled me into her wide lap and held me, rocking, as I wept.

Fast Forward.

Three weeks after my father died, in May 2022, I flew to Bimini, in the Bahamas, to join a group of people swimming with dolphins. Although the trip was a spontaneous decision, it was also something I'd been dreaming about, ever since those years in high school when we'd studied John Lilly's work with dolphin communication, and when I'd listened to Roger and Katy Payne's recordings of whale-song. Sonic seeds planted in my memory, rediscovered decades later.

As I booked my flights and paid in advance for the excursion, I hadn't known my father was dying, although some part of me was aware—every single time I saw him—that it might be the last. I wasn't expecting this to be some sort of echo of that Mexico journey I'd made following the death of my mother. But there I was, exactly three weeks past losing my second parent. Flying south.

Having spent the night at a hotel near the airport, it was still early in the morning when I took my seat on a seaplane in Fort Lauderdale, and only forty-five minutes later splash-landed in the stunning turquoise waters of Bimini. Within a couple of hours, a dozen of us were boarding the WildQuest catamaran, stashing our fins and snorkels where they'd be easily accessible. Drenched in grief, yet also transfixed by the breathtaking Caribbean Sea, I sat close to the bow and watched the colors shift and merge. Conversations flowed around my back and over my head. I stayed aware enough to catch the basic guidelines for what we would do if and when any of the dolphin pods were spotted by the crew—which consisted of a jolly, Italian-accented captain, two deeply tanned young men, one bikini-clad woman, and another fair-skinned woman wearing a long-sleeved rash guard. The four crew members were stationed at various positions, their binoculars scanning the horizon in all directions.

I had specifically chosen to join a trip with this company because its aim is toward encounters with dolphins in the wild, a practice they've been cultivating for more than twenty years. For this reason, there was no guarantee we would get to see the dolphins, much less get to experience them in the water.

"It's entirely up to the dolphins to be visible to us or not," explained Kathleen, the woman with long sleeves. "Because we've been out here for so many seasons, they know us and recognize us, but the ocean is huge. They can stay entirely hidden if they want to."

Lean, freckled, and knowledgeable, Kathleen shared a cascade of useful details about dolphin behavior and what we could expect. For five days in a row, we would be spending seven or eight hours at a stretch on the catamaran, searching the open waters around Bimini where the pods have often been encountered. We might have the chance to see Atlantic spotted dolphins, spinners, bottlenose—or none of them at all.

"We will do our best!" promised the Italian captain.

I allowed myself to hope for at least one chance to connect with a dolphin underwater. And moments later, the crew announced that a pod of Atlantic spotted dolphins had been sighted not far away. As the captain steered us in the direction of the dolphins, whose sleek bodies were clearly moving closer to us as we approached them, we were told to grab our gear and prepare to get into the water. I reached with a gleeful thrill for my borrowed mask and snorkel, my bright-yellow fins. More directives were hastily provided as the catamaran came to a stop and the boat bobbed for a few minutes in the stunningly clear ocean.

"Keep your arms at your sides; do not reach out to try to touch the dolphins; stay as calm as possible, especially as you get into the water; and . . . here they are."

It was challenging to wait my turn as each person was gently guided off the pair of ladders at the stern of the catamaran. Although I felt a rush of adrenalin pouring through my body the moment I entered the sea, I also sensed the soothing comfort of its warm-enough temperature. Following Kathleen's lead, I took a deep breath and made a surface dive to see what was happening below, reassured by the fact that I could see the white sandy bottom maybe fifteen feet down.

I had already been having a series of what might be called otherworldly experiences since a couple of days *before* my father's death—starting, improbably enough, in the parking lot of the suburban shopping center in Schenectady where I had gone to pick up his morphine drops. Climbing out of my rental car, I saw a sunset so dazzling, so utterly unlike anything I had ever witnessed in my hometown, that the vision took my breath away. Streaks of light pointed in a cinematic fan shape spreading up and out from the horizon. Everything around me went quiet and vivid and luminous.

"*God?*" I gasped out loud. "*Is that you??*"

Nobody else seemed aware in the least. People pushed their shopping carts and went about their business-as-usual while I stood and gaped at the miraculous sky show. Later that same day, I stood in my father's kitchen, eating an orange whose tart juiciness made every cell of my mouth come alive. These weren't out-of-body moments. They were *in-body* moments. Ecstatic moments.

When I tried recounting these experiences to a friend a few weeks later, he said, "Oh, of course. You were in the Death Zone."

As many others have written, grief was cracking me open—even though I guess that for some people, it seems to slam them shut. Not unlike that window of time after my mother's death, I saw and felt and heard what might otherwise have stayed obscured by my thoughts, interpretations, fears. Instead, what I felt was a kind of porousness to everything, not the usual semipermeability of my senses, filtered with vigilance and self-protection, but rather a new nakedness—with or without my having chosen it. Whatever you might call it, that zone carried me into the astonishing waters of Bimini, where the dolphins were waiting.

Holding my breath during a dive, I floated in a state of pure awe as these fellow mammals swam in every direction, toward and away from me. I watched Kathleen—former modern dancer (as I later learned)—while she moved effortlessly in graceful arabesques that matched the dolphins' shapes. A few of the dolphins made eye contact with me, seemed to be smiling, and nodded their heads as if in wordless agreement about a shared mystery. I returned to the surface so I could observe for longer as I breathed through my snorkel. A group of four dolphins hung in vertical position near me; another pair touched flippers and swam in sync as though they were holding hands.

I have no idea how much time passed before the pod departed and we all climbed back aboard the boat. I was in a giddy electric bliss, my entire being intoxicated by waves of gratitude and grace. I couldn't speak to anyone, couldn't stop the joyful tears streaming down my face and mixing with the salt water on my skin. And then, maybe twenty minutes later, a crew member called out that the pod had returned; we could get back into the water if we wished.

"There's no way of knowing what the other days of the week might bring," he encouraged. "You might want to take this opportunity."

On with my fins and mask and snorkel, I was among those who got right back in, *yes please*. This departure and return of the dolphins happened three more times over the course of a couple of hours. Kathleen commented that this behavior wasn't typical. "You got some extra blessings," she said.

And then the pod stayed gone.

As it turned out, the rest of the week was nothing like that first day. The dolphins kept their distance from the boat—close enough for us to see their fins cutting through the surface of the water, and their occasional leaps into the air, but not approaching us the way they had before. Nor was the choppy ocean calm enough for the captain to drop anchor. In our hours and days of motoring and sailing, we saw some manta rays and some turtles. We snorkeled around the edges of a submerged shipwreck and spent time free diving in an area that was referred to as "The Road to Atlantis," with wide, flat stones that seemed to be arranged like an underwater highway on the sandy bottom. Divers came from all around the world to visit that place, we were told.

As for the dolphins, it was just that one day, the one encounter I'd allowed myself to hope for. I'd been given enough, more than enough.

ECHOES FROM THE TREE OF LIFE

IN THE YEAR 2010, FOLLOWING THE DEATH OF HIS COUSIN from cancer, a man named Itaru Sasaki set up a disconnected telephone booth—with a rotary-style phone—in his garden in the small town of Otsuchi, Japan, thus creating what he called 風の電話, *kaze no denwa*, known in English as the "phone of the wind." In describing its purpose, he acknowledged that it was not intended as an "earthly system" of staying in touch with his cousin, but as a source of comfort in his mourning.

Just one year later, an earthquake of 9.1 magnitude was followed by a thirty-foot tsunami, wreaking massive destruction along the northeastern Japanese coast and causing thousands of deaths. While the earthquake and subsequent tsunami's impact on the nuclear power plant in Fukushima claimed, understandably, the spotlight of scrutiny, the town of Otsuchi recorded the highest number of missing persons. And the "phone of the wind" became a place of pilgrimage for residents seeking to connect with their lost loved ones.

"If you're out there, please listen to me . . ." begins one subtitled YouTube video depicting the project.[1] We viewers are eavesdroppers, listening in on some of the most poignant one-way conversations with the air.

A 2021 documentary about the legacy of the tsunami made by a French-German film crew includes extended footage of people using the Otsuchi phone box. “I miss your voices,” a gray-haired man speaks tearfully into the mouthpiece of the old black telephone, the silent receiver cupping his ear. “I’m so sorry I didn’t get to say goodbye,” says a young woman, after calling her grandfather’s number.

A young man who identifies himself as a telecom engineer explains to the camera that cable connections promise that “there’s always someone to talk to.” When the cable isn’t attached, however, “you are connected to something invisible, and that’s why you are able to communicate with the deceased.”

“It was good to speak directly to him,” another woman tells the film’s off-screen interviewer, after concluding her own few minutes inside the booth. “It’s like he was listening to me.”[2]

In recent years, especially in response to the daunting cascade of death from the COVID-19 pandemic, hundreds of similar “wind telephones” have been established around the world, from Australia to Ireland and everywhere in between. They can be found in most states throughout the United States, with continually updated locations tracked by a website called mywindphone.com.

. . .

My father’s funeral in April 2022 was delayed by the eight days of Passover. The complicated rules of Judaism require that no one can be buried during Passover, because the official mourning period of *shiva* cannot begin until the holiday has ended.

I had a hard time making sense of all this, despite the Rabbi’s explanation that *shiva* rituals are undertaken as a community practice; therefore, the bereaved have to wait so that friends and synagogue members can properly attend to the needs of the family without feeling conflicted about focusing on their Passover celebrations. A long-lost memory returned when I heard this, something Mr. Friend had taught us all those decades back in Hebrew school: if a funeral and a wedding procession meet at a crossroads, the funeral should wait so that the wedding can proceed first. Because joy takes precedence over grief.

During the week of *shiva* for my mother, I chose to break the rules—about not leaving the house—because I needed to go swimming. Immersing in water has always been my antidepressant, my meditation, my solace. I also needed extra consolation in private because on the nights of *sitting shiva* for her, the men and women of the Orthodox synagogue community had placed themselves on separate sides of our living room. Those gender rules that I fled from so many years ago were the same ones I believed she had been refusing whenever she went shopping on Shabbat mornings instead of sitting on the side with all the women. *She would not have liked this*, I fumed silently—for her and for myself.

The night before my father's funeral, the weather turned sharply cold. Even though it was late April, several inches of snow covered everything, even the budding spring crocuses. At the cemetery, we were bundled back into our winter coats, layered with scarves and hats and boots. As our family and friends arrived and the funeral home director guided people to their seats, the clouds cleared until the sky was bright blue and sharp, fresh with the scent of wet pine.

Some part of me was recalling my mother's funeral more than twenty years earlier. The Rabbi had taken the four of us aside to perform what's called the *k'riah*; he explained that for a mourning spouse the tearing of garments is on the right side of the chest, while for children the rip symbolizes the severing of connection through blood, and therefore must take place on the left side, over the heart. The cloth of my blouse resisted a little before it yielded, and I sobbed at the sound.

This time, for my father, the Rabbi stood with the three of us, and he read a poem about what the *k'riah* means, originally written by Rabbi Harold M. Schulweis. The last two stanzas are:

> Each of us a letter in
> the Torah scroll
> Together our lives are intertwined
>
> Our common fate and faith
> our common destiny
> find us like the stiches of the parchment
> when any of us is lost
> The holy text is torn.
> In memory we are mended.[3]

At the graveside, in addition to the Rabbi, my siblings and I each offered a eulogy, and so did my father's eldest grandson Ezra. I kept flashing back to the long-ago eulogy for my mother that I had written and read aloud. Even though I tucked the notes into the pocket of my coat, when we returned from the cemetery the papers had vanished. I never found them.

Now, as the ice-capped trees dripped onto the fresh snow and ghosts of steam rose from the nearby stones, our cousin Moshe stood up. Having left Israel for a life in New York City, where he married and had children, Moshe was the youngest son of my Uncle Joseph, who survived Buchenwald with my father. My cousin had always been very close to my father, closer in many ways than he was to his own parents back in Israel; he often reached out to my father to ask for advice about running his small business, and he visited often. Moshe, his cheeks flushed from the cold, explained that he wanted to chant a special prayer. He needed no book, of course, having been raised with ultra-Orthodox traditions and knowing so much by heart.

I didn't recognize the prayer or its old-world melody, but I felt Moshe's sorrowful voice echo across all the ancestral generations and resonate toward the heavens. Then a gust of wind blew up out of nowhere, grabbing one of the Rabbi's pages—containing the *k'riah* poem—and floating it down into my father's grave.

. . .

"Quantum Listening—which flows out of Deep Listening—pays attention to multiple realities," wrote Pauline Oliveros. "It's 'perception at the edge of the new. Jumping like an atom out of orbit to a new orbit—creating a new orbit—as an atom occupies both spaces at once one listens in both places at once.' The Quantum Listener 'listens to listening.'"[4]

. . .

An orphan is another kind of island. I fly back to California from Florida, after leaving Bimini and the dolphins, on a date that marks exactly one month since my father's death. From my window seat on the plane, staring at the expanse of sky and clouds and chessboard landscapes, I lean my forehead against the window. It occurs to me that I am once again flying far away from my childhood—except that from now on, there will be no parent waiting for me to return home anymore.

Although hearing takes place immediately and automatically, listening is always an interpretation. Sometimes the brain only processes what we have heard afterward, much later, in a delayed time.

Replaying in my mind the way all of us took turns shoveling the frozen clods of dirt onto my father's plain wooden casket, I hear the repetitive finality of that sound. And that is when I sense with new inner awareness what happened in our last, almost-wordless conversation.

My father gave three squeezes to my hand. Distinct and strong and silent and clear.

I—Love—You.

I listen again, deeper.

. . .

"Hello from the children of planet Earth," says the voice of the English greeting, one among the fifty-five languages recorded on the NASA Voyager's Golden Record, beginning with Akkadian, spoken in Sumer six thousand years ago, and ending with Wu, a modern Chinese dialect. Not a coincidence that these vocabularies attempt to represent the variety as well as the millennial stretches of time we humans have been talking and listening to each other. After all, "it will be forty thousand years before [the Voyager probes] make a close approach to any other planetary system."[5]

Describing the launching of the spacecraft with the contents of the Golden Record as a "bottle into the cosmic ocean," astronomer Carl Sagan suggested it "says something very hopeful about life on this planet."[6]

On Valentine's Day in 1990, "after the spacecraft had passed the orbits of Neptune and Pluto, the cameras of Voyager 1 pointed back toward the sun and took a series of pictures of the sun and the planets, making the first ever 'portrait' of our solar system as seen from the outside."[7] It's been written that Carl Sagan was the one who insisted that the cameras be turned around to focus on our solar system before *Voyager 1* departed altogether. "Although it had completed its incredible mission to photograph Saturn and Jupiter, and their moons, it managed to capture Earth as a single pixel in the center of scattered light rays."[8]

. . .

In this third decade of the twenty-first century, when many of us seem to be finding it harder and harder to listen deeply to each other—whether for political reasons or the sheer noise from every direction—it seems both more dispiriting and more necessary to turn our ears together toward the sounds of the natural world. Maybe this turning of our collective attention can bring us a new variation of what happened back on Christmas Eve in 1968, when the *Apollo 8* astronauts orbiting the moon didn't just beam back images of our shared, fragile planet, but also allowed us, together, to listen to them taking turns reading the first ten verses from the book of Genesis.

> "We were told that on Christmas Eve we would have the largest audience that had ever listened to a human voice," recalled [Frank] Borman during 40th anniversary celebrations in 2008. "And the only instructions that we got from NASA were to do something appropriate." [. . .]
>
> The mission was also famous for the iconic "Earthrise" image, snapped by [Bill] Anders, which would give humankind a new perspective on their

home planet. Anders has said that despite all the training and preparation for an exploration of the moon, the astronauts ended up discovering Earth.[9]

. . .

Roger Payne died in June 2023. Fellow biologists and conservationists credit his whale-song recordings with raising empathic awareness leading to environmental impacts almost immeasurable in their scope. What became an anthem-like soundtrack for the Save the Whales movement meant that several whale species were rescued from extinction.

In an essay published in *Time* magazine just days before his death at the age of eighty-eight, Payne described his current efforts with Project CETI, the Cetacean Translation Initiative, which focuses on "using advanced machine learning and state-of-the-art robotics to listen to and translate the communication of sperm whales."

Insisting that our *collective* existence hangs in the balance, he said: "I believe that awe-inspiring life-forms like whales can focus human minds on the urgency of ceasing our destruction of the wild world. Many of humanity's most intractable problems are caused by disregarding the voices of the Other—including non-humans."[10]

. . .

In her *New Yorker* article about current explorations involving interconnections between human and animal perception, Elizabeth

Kolbert writes: "At the time the Voyagers set out [in 1977], no one knew what, if anything, the humpbacks were trying to convey. Today, the probes are more than ten billion miles from Earth, and still no one knows. But people keep hoping."[11]

. . .

Dr. Charles Limb, the inner ear surgeon studying improvisation and the brain, is also the specialist monitoring the progress of my *vestibular schwannoma*, sometimes called an *acoustic neuroma*. He informs me that using the word "benign" is not entirely adequate to the task—because in this case, although the tumor is benign in the sense of not being cancerous, it's not benign in the sense of its location.

"Do you want to see your MRI?" he asks when we meet via Zoom for our first appointment. When I say yes, he brings it up on the screen and points to the area under observation. He explains that my facial nerve is very nearby—the nerve that controls the movement of my face—and also nearby is my brain stem.

"We really don't want it to grow anywhere near either of those," he says.

No, we do not.

I can sense myself leaning toward the computer screen, that child inside me still listening with my entire body, listening *hard*. Yet I am also sitting very still. Because I'm not just hearing his words but trying to absorb their meaning, while simultaneously trying to comprehend something about the interior map of my skull, the specific universe we are looking at together.

Dr. Limb wants me to get a second MRI so that we can see what changes have taken place since the scan I had six months earlier. This information will be essential in deciding what, if anything, to do next. Surgery is one option; targeted radiation is another. "Watch and wait" is a third.

"The thing is," Dr. Limb adds, "if you're going to have a brain tumor, this is the kind of brain tumor to have."

"Did you just say I have a brain tumor?"

He did.

Thankfully, my second MRI shows "no change," which means that we can watch and wait. That is when I find out that Dr. Limb is a jazz saxophonist studying brain activity during improvisational music-making.

> In one [of Dr. Limb's experiments], he found that improvising musicians showed: 1) deactivation of the dorsolateral prefrontal cortex, which among other functions acts as a kind of self-censor, and 2) greater activation of the medial prefrontal cortex, which connects to a brain system called the "default network." The default network is associated with introspective tasks such as retrieving personal memories and daydreaming. It has to do with one's sense of self.[12]

I keep wishing I could find a way to turn inside out to observe the intricate nerves and semicircular canals of my inner ears. When I took a few opera lessons while doing research for my first novel, my teacher recorded me singing so I could listen to myself from the outside, the way she was listening. Now I'm thinking about the way that when I cup my hand over my right ear, I hear my voice extra loudly *inside* my head.

Can you hear me now?

I'm here, I'm here.

. . .

"An accent is not just about pronouncing words," writes Yao Xiao, "it's a way of being, a posture of life." She goes on to describe the way that listening to her father speaking in his native dialect, one that she herself never learned, is "like hearing warmth itself. If the sun made sounds when it moved slowly across the sky, perhaps it would sound as flowing, musical and thunderous as it did in Shanxi."[13]

. . .

On July 4, 2023, my mother would have been celebrating her ninety-third birthday. She has been gone from the world for nearly twenty-three years, and despite all the efforts of my complicated brain, somehow this fact continues to seem untranslatable. I'm so grateful that my nephew digitized her USC Shoah Foundation testimony, the recordings made using now-obsolete VCR technology. *You don't know Pushkin?* I float again in the memory of her symphonic mother tongue.

. . .

I keep thinking about the "phone of the wind," about how we long for traces of those we have lost. Maybe we need the sonic equivalent of someone's hand squeezing ours. We search for a hum in outer space or underwater; we detect a whisper in the forest; we chant from sources of ancient music helping us to feel less alone.

. . .

A few months ago, I stood before a very old stone carving that my friends placed on their lush property, located on the North Shore of the island of Kauai. The garden island. When Steven explained to me that the image represented the Tree of Life, I felt a rush of revelation so strong it took my breath away. Then I started singing the prayer from my childhood in synagogue, the one whose harmony always lifted my pained heart. I knew that the tree was a metaphor referring to the Hebrew Bible, that the words referenced Genesis, and I'd always seen the translated English as "it is a tree of life . . ." But for the first time I noticed something even more moving to my ears than the tender beauty of the melody. The sacred text of the Torah was being given a feminine pronoun. *Eitz chaim hee*—as in, *she* is a tree of life.

עֵץ־חַיִּים הִיא לַמַּחֲזִיקִים בָּהּ וְתֹמְכֶיהָ מְאֻשָּׁר׃ {פ}

She is a tree of life to those who grasp her,
And whoever holds on to her is happy.

ACKNOWLEDGMENTS

I've been graced with an extended web of generous friends and family and colleagues. Individually and collectively, they sustain me with writerly essentials of companionship and nourishment and meaningful connection. Sometimes they have healed me with meals or therapy or medicine; they have hosted me in guest cottages and offices; they have provided official and unofficial artist residencies; and a few have read or listened to pages while this work was taking shape. In sum, we have shared countless rooms and portals both real and virtual.

I would not have finished this book (or any of my books) without the support and kindness of many who—for various reasons—are not mentioned below. This is especially true with my nonfiction, since sources of material include personal interviews and research assistance and storytelling and anonymity. That being said, I want to thank:

Ahria Wolf, Alan McEvoy, Alan Ritterband, Alison Owings, Alison Wearing, Amy Cooke, Amy Ferris, Anandra George, Ana Thiel, André Salvage, Anita Barrows, Anna Deavere Smith, Anne Germanacos, Anne Peled, Annette Wells, Ann Randolph, Ash Verjee, Barbara and Kim Marienthal, Barbara Yoder, Belinda Lyons-Newman and Dan Newman, Beverly Donofrio, Bonnie Tsui, Brendan O'Brien, Carol Haggas, Casper Calderola, Catherine Arduini (of blessed memory), Cathi Colas, Cathy

Fischer, Charlene Dyer, Charles Benner, City of Berkeley Civic Arts Grants, Claire Ellis and Chuck Greenberg, Claudia Tellez Campos, Clayre Haft, CMarie Fuhrman, Corbin Smith, David Decker, David Dinner & Kim Delgado, David Rosner, Debbie and Jay Yablon, Debora Hoffman, Delisa Sage, Denise Kaufman, Devi Jacobs, Devi Laskar, Diane Walder, Edie Meidav, Eleonora Isunza, Elisabeth Malkin, Elizabeth Kert, Elizabeth Garfield, Ellen Comisar, Enrique Vallejo, Eric Slayton and Elena Lyakir, Erin O'Malley, Eve Hersov, Ezra Brettler, Frances Dinkelspiel, Fundación Valparaiso, Garrett Smith, George Goldsmith, Ginny Rorby, Gloria Saltzman, Gobi Stromberg, Heather Hiett, Hedy Roma and Tony Gonzales, Helena Haase, Helen Epstein, Indigo Moor, Irina Posner, Jacki Lyden, Janet Dawson and Doug Clarke, Janice Ayers, Jason Craige Harris, Jennifer Rosner, Jill Culver, John VanDuyl (of blessed memory), Jozie Rabyor, Judy Levin, Julia McNeal, Julia Sattler, Julie Freestone and Rudi Raab, Julie Robinson, Karen Bender, Kathi Jordan, Kathy Barr, Katy Butler, Krista Comer, Kristen Millares Young, Laura Miñano Mañero, Laurie Wagner, Lenore Pernaul, Leslie Wahlquist, Lissy Abraham, Lola and Andrew Fraknoi, Lori Saltzman & Kathy Altman, Luisa Giulianetti, Lynn Eve Komaromi, Lynn Newcomb, Magda Bogin & Under the Volcano, Marge and Bob Berger, Marian Palaia, Marie-France Rosner, Mark and Ximena Martel, Mary Ford and Rob Lewis, Meg Corman, Meredith Sabini, Meta Pasternak, Michael Singer (of blessed memory), Micki and Lee Evslin, Moshe Rosner, Nadia Brown, Nancy Borris, Nina Wise, Pamela Gaye Walker and John Walker, Pattie Hogan, Patty Joslyn & Larry Babic, Pat Wood, Paula Whyman, Priya Parmar, Rachael Cerrotti, Rachael McDonnell, Raphael and Saint Rosner, Randall Alifano, Randy Susan Meyers, Rene Denfeld, Richard Zimler, Robert Ward, Robin Read, Rona and Mick Renner, Sandra Lorenzano, Sara Glaser, Sarah Kobrinsky, Sarajane Giddings, Sari Levine,

Sheri Shuster, Stefa Seltzer, Stephanie Flom, Steven Ruddell and Merlyn Wenner, Stuart and Deb Fiedler, Susan Griffin, Susan Hall and Steve McKinney, Susanne Pari, Suzanne LaFetra, Terry Lowe, the Community of Writers, the Mesa Refuge, Tony LeHoven, Tora and Kirk Smart, Trish and Rigdon Currie, Troy Gangle, Valerie Mejer, Vijaya Nagarajan, Walter Gruenzweig and Brunhild Foelsch, Yaffa Lown, Yonit Hoffman, Zdena Berger (of blessed memory).

Thank you to my terrific agent Miriam Altshuler. Thank you to the dazzlingly talented team at Counterpoint—especially to Nicole Caputo for designing a gorgeous cover; to Rachel Fershleiser and Lena Moses-Schmitt and Megan Fishmann for champion-level marketing and publicity; to production editor Laura Berry, copy editor Colin Legerton, and proofreader Madelyn Lindquist for meticulous attention to detail; to production manager Olenka Burgess for the elegant interior design; to Jack Shoemaker for founding and sustaining such an invaluable publishing endeavor. Thank you (meager words!) to my forever editor Dan Smetanka.

Thank you to *all* of my students, then and now.

Last but not least, thank you, reader, for listening to these words—including those quiet spaces between, above, and below.

NOTES

PORTAL, AN INTRODUCTION

1. Theodor Reik, *Listening with the Third Ear: The Inner Experience of a Psychoanalyst* (Garden City, NY: Garden City Books, 1948), 145.
2. Jens Korff, "Deep Listening (Dadirri)," Creative Spirits, March 18, 2023, www.creativespirits.info/aboriginalculture/education/deep-listening-dadirri.

FIRST SOUNDS

1. Anne Karpf, *The Human Voice: How This Extraordinary Instrument Reveals Essential Clues About Who We Are* (New York: Bloomsbury Publishing, 2006), 27.
2. Karpf, *Human Voice*, 72.
3. Karpf, *Human Voice*, 27.
4. Kristin Wong, "What Lullabies Teach Us About Language," *Catapult*, August 4, 2022, catapult.co/stories/what-lullabies-teach-us-about-language-motherhood-babies-kristin-wong.
5. Alok Jha, "Babies Remember Melodies Heard in Womb, Study Suggests," *The Guardian*, October 30, 2013, www.theguardian.com/science/2013/oct/30/babies-remember-melodies-womb-study.
6. Sally Goddard Blythe, *The Genius of Natural Childhood: Secrets of Thriving Children* (Gloucestershire, UK: Hawthorn Press, 2011).
7. Jenny Marder, "Why Are So Many Lullabies Also Murder Ballads?," *PBS NewsHour*, August 13, 2014, www.pbs.org/newshour/science/many-lullabies-murder-ballads.
8. Karpf, *Human Voice*, 73.
9. Linda O'Brien, "An Exploration of How the Use of the Couch in Psychoanalysis May Impact the Experience of Analysis for the Patient" (thesis, Dublin Business School, School of Arts, May 2019), esource.dbs.ie/bitstream/handle/10788/3778/hdip_obrien_l_2019.pdf.
10. "Dolphins Use a Baby Voice to Talk to Their Young," *BBC Newsround*, June 28, 2023, www.bbc.co.uk/newsround/66029015.

11. Stefanie Faye, "The Importance of Sound and How It Affects Our Wellbeing," September 17, 2019, in *Mindset Neuroscience*, podcast, season 1, episode 6, 36:02, stefaniefaye.com/podcast/importance-of-sound-mnt-006/.
12. Wong, "Lullabies."
13. Dr. Viorica Marian, phone interview with author, July 8, 2019.
14. Sayuri Hayakawa and Viorica Marian, "How Language Shapes the Brain," *Scientific American*, April 30, 2019, blogs.scientificamerican.com/observations/how-language-shapes-the-brain/.
15. Dr. Viorica Marian, phone interview with author, July 8, 2019.
16. Thomas Lewis, Fari Amini, and Richard Lannon, *A General Theory of Love* (New York: Vintage Books, 2000), quoted in Maria Popova, "Relationship Rupture and the Limbic System: The Physiology of Abandonment and Separation," *The Marginalian*, September 6, 2022, www.themarginalian.org/2022/09/06/general-theory-of-love-separation/.
17. Lewis, Amini, and Lannon, *General Theory of Love*, 76.
18. Lewis, Amini, and Lannon, *General Theory of Love*, quoted in Popova, "Relationship Rupture."
19. Molly Rains, "Crocodiles Are Alarmingly Attuned to the Cries of Human Infants," *Science*, August 8, 2023, www.science.org/content/article/crocodiles-alarmingly-attuned-cries-of-human-infants.
20. "Bilingualism as a Natural Therapy for Autistic Children," Neuroscience News, June 3, 2021, neurosciencenews.com/asd-bilingual-18551/.

LIVING IN THE SOUNDSCAPE

1. R. Murray Schafer, "Open Ears" (lecture, Acoustic Ecology: An International Symposium, Melbourne, Australia, March 2003).
2. R. Murray Schafer, *The Soundscape: Our Sonic Environment and the Tuning of the World* (Rochester, VT: Destiny Books, 1993), 43.
3. William Duckworth, *Talking Music: Conversations with John Cage, Philip Glass, Laurie Anderson, and Five Generations of American Experimental Composers* (Boston, MA: Da Capo Press, 1999), 163.
4. "Pauline Oliveros (1932-2016)," The Center for Deep Listening, www.deeplistening.rpi.edu/deep-listening/pauline-oliveros/.
5. Andreas Weber, *The Biology of Wonder: Aliveness, Feeling, and the Metamorphosis of Science* (Gabriola Island, BC: New Society Publishers, 2016), 31.
6. Weber, *Biology of Wonder*, 160.
7. Daisy Hildyard, "War on the Air: Ecologies of Disaster," *Emergence Magazine*, June 27, 2022, emergencemagazine.org/essay/war-on-the-air/.
8. *The Year Earth Changed*, directed by Tom Beard (Bristol: BBC Natural History Unit, 2021), 48 min.

9. Adam Loften and Emmanuel Vaughan-Lee, "Sanctuaries of Silence: An Immersive Listening Journey into One of the Quietest Places in North America," Global Oneness Project, www.globalonenessproject.org/library/virtual-reality/sanctuaries-silence.
10. Gertrude Stein, *How to Write* (Paris: Plain Edition, 1931).
11. Seth S. Horowitz, "The Science and Art of Listening," Opinion, *New York Times*, November 9, 2012, www.nytimes.com/2012/11/11/opinion/sunday/why-listening-is-so-much-more-than-hearing.html.
12. Faye, "Importance of Sound."
13. Douglas Heingartner, "Now Hear This, Quickly," *New York Times*, October 2, 2003, www.nytimes.com/2003/10/02/technology/now-hear-this-quickly.html.
14. Lauren Murrow, "Speed-Listening and the Trouble with 'Podfasters,'" *Wired*, September 29, 2018, www.wired.com/story/stop-speed-listening-podcasts/.
15. David George Haskell, *Sounds Wild and Broken: Sonic Marvels, Evolution's Creativity, and the Crisis of Sensory Extinction* (New York: Penguin, 2023), 18.
16. Susan Hall, "The Intimacy of Loneliness," *Blog: Musings from the Studio*, September 2023, www.susanhallart.com/blogs.
17. "Sounds You Don't Hear Anymore," Slow Radio, BBC Radio 3, www.bbc.co.uk/programmes/articles/2g7BNYczrSb5cWQd0z5WHC1/sounds-you-dont-hear-anymore.
18. Oliver Sacks, *Musicophilia: Tales of Music and the Brain* (New York: Knopf, 2007), 188, 195.
19. *If These Walls Could Sing*, directed by Mary McCartney (London: Abbey Road Studios, 2022), 86 min.
20. Dr. Viorica Marian, phone interview with author, July 8, 2019.
21. Valerie Brown, "Leaders: 8 Steps to Be a Mindful Listener," *Education Week*, May 9, 2016, www.edweek.org/education/opinion-leaders-8-steps-to-be-a-mindful-listener/2016/05.
22. Bernie Krause, *Voices of the Wild: Animal Songs, Human Din, and the Call to Save Natural Soundscapes* (New Haven, CT: Yale University Press, 2015), 77.
23. Hall, "Intimacy of Loneliness."

WHISPERS AND HEALING

1. Jessica Weinberger, "A Brief History of Therapy," Talkspace, February 24, 2020, www.talkspace.com/blog/psychotherapy-history-of-therapy/.
2. Katherine Harvey, "Confession as Therapy in the Middle Ages," Wellcome Collection, February 20, 2018, wellcomecollection.org/articles/WovlRioAAHW6Xfqn.

3. Stephen Nachmanovitch, *The Art of Is: Improvising as a Way of Life* (Novato, CA: New World Library, 2019), 41.
4. Encyclopedia.com, s.v. "Evenly-Suspended Attention," www.encyclopedia.com/psychology/dictionaries-thesauruses-pictures-and-press-releases/evenly-suspended-attention.
5. Kendra Cherry, "Anna O's Life and Impact on Psychology," Verywell Mind, September 13, 2023, www.verywellmind.com/who-was-anna-o-2795857.
6. John Launer, "Anna O and the 'Talking Cure,'" *QJM: An International Journal of Medicine* 98, no. 6 (June 2005): 465–66, doi.org/10.1093/qjmed/hci068.
7. James E. Miller, *The Art of Listening in a Healing Way* (Fort Wayne, IN: Willowgreen Publishing, 2003).
8. Rachel Kurzius, "How Zoom Killed the Fine Art of Interrupting," *Washington Post*, May 10, 2021, www.washingtonpost.com/lifestyle/magazine/video-calls-unnatural-conversation/2021/05/06/fb2cdca4-a203-11eb-85fc-06664ff4489d_story.html
9. Kurzius, "Zoom."
10. Maria Montessori, quoted by Cleary Vaughan-Lee, executive director of Global Oneness Project (email newsletter, December 18, 2021).
11. "What Is a Horse Whisperer? The Ins and Outs of a Horse Psychologist," Horse Riding Guide, www.horseridingguide.com/horse-whisperer/.
12. Alysa Landry, "Jay Begaye: Navajo Horse Whisperer," *ICT News*, last modified September 13, 2018, ictnews.org/archive/jay-begaye-navajo-horse-whisper.
13. Marina Sarris, "Something about a Horse: Finding Benefits for Autism in Therapeutic Riding," SPARK, September 26, 2022, sparkforautism.org/discover_article/something-about-a-horse-finding-benefits-for-autism-in-therapeutic-riding/; and Robin L. Gabriels et al., "Randomized Controlled Trial of Therapeutic Horseback Riding in Children and Adolescents with Autism Spectrum Disorder," *Journal of the American Academy of Child and Adolescent Psychiatry* 54, no. 7 (2015): 541–49, doi.org/10.1016/j.jaac.2015.04.007.
14. David Kelly, "Children with Autism Spectrum See Benefits from Equine Therapy," CU Anschutz Medical Campus, University of Colorado, October 30, 2018, news.cuanschutz.edu/news-stories/children-with-autism-spectrum-have-immediate-and-long-term-benefits-from-therapeutic-horseback-riding-researchers-show.
15. Martha Brennan, "The Horse Boy Method: Horseman on a Mission,"

Irish Examiner, August 19, 2022, www.irishexaminer.com/lifestyle/health andwellbeing/arid-40943144.html.

16. Karpf, *Human Voice*, 17.
17. Lydia Denworth, "Brain Waves Synchronize When People Interact," *Scientific American*, July 1, 2023, www.scientificamerican.com/article /brain-waves-synchronize-when-people-interact/.
18. Robert Martone, "Music Synchronizes the Brains of Performers and Their Audience," *Scientific American*, June 2, 2020, www.scientificamerican.com /article/music-synchronizes-the-brains-of-performers-and-their-audience/.
19. Anna Deavere Smith, *Talk to Me* (New York: Random House, 2000), 39.
20. Anna Deavere Smith, interview with author, March 19, 2021.
21. Anna Deavere Smith, "Some Notes on *Notes from the Field*," *Literary Hub*, May 21, 2019, lithub.com/anna-deavere-smith-some-notes-on-notes-from -the-field/.
22. Preeti Vankar, "Mental Health Treatment or Counseling among US Adults 2002–2021," Statista, November 29, 2023, www.statista.com/statistics /794027/mental-health-treatment-counseling-past-year-us-adults/.
23. Holly A. Swartz, "The Role of Psychotherapy During the COVID-19 Pandemic," *American Journal of Psychotherapy* 73, no. 2 (June 1, 2020): 41–42, doi.org/10.1176/appi.psychotherapy.20200015.
24. World Health Organization, "COVID-19 Pandemic Triggers 25% Increase in Prevalence of Anxiety and Depression Worldwide," news release, March 2, 2022, www.who.int/news/item/02-03-2022-covid-19-pan demic-triggers-25-increase-in-prevalence-of-anxiety-and-depression-world wide.
25. Michael Stadler et al., "Remote Psychotherapy during the COVID-19 Pandemic: A Mixed-Methods Study on the Changes Experienced by Austrian Psychotherapists." *Life* 13, no. 2 (January 29, 2023): 360, doi .org/10.3390/life13020360.
26. Terry Tempest Williams, *When Women Were Birds: Fifty-four Variations on Voice* (New York: Macmillan, 2013), 50.

THE SPACES BETWEEN

1. David Rothenberg, *Nightingales in Berlin: Searching for the Perfect Sound* (Chicago: University of Chicago Press, 2019), 18.
2. Williams, *When Women Were Birds*, 27.
3. Peter Reuell, "Muting the Mozart Effect," *Harvard Gazette*, December 11, 2013, news.harvard.edu/gazette/story/2013/12/muting-the-mozart-effect/.
4. James Wood, "The Graceful Rebellions of Wolfgang Amadeus Mozart,"

New Yorker, May 22, 2023, www.newyorker.com/magazine/2023/05/29/mozart-in-motion-his-work-and-his-world-in-pieces-patrick-mackie-book-review.

5. Weber, *Biology of Wonder*, 210.
6. Reuell, "Muting the Mozart Effect."
7. Jaron Lanier, "What My Musical Instruments Have Taught Me," *New Yorker*, July 22, 2023, www.newyorker.com/culture/the-weekend-essay/what-my-musical-instruments-have-taught-me.
8. Grace Slick (@GraceSlick_JA), "I wrote White Rabbit on a red upright piano that cost me about $50. It had eight or 10 keys missing, but that was OK because I could hear in my head the notes that weren't there," Twitter, September 26, 2022, twitter.com/GraceSlick_JA/status/1574413987059572736.
9. Keith Goetzman, "Keith Jarrett: How I Create," *Utne Reader*, October 9, 2007, www.utne.com/arts/howicreate.
10. Will Hermes, "The Story of '4'33",'" National Public Radio, May 8, 2000, www.npr.org/2000/05/08/1073885/4-33.
11. Williams, *When Women Were Birds*, 65.
12. Nadia Sirota, "The Space Between," WQXR, August 15, 2011, www.wqxr.org/story/152650-the-space-between/.
13. Goetzman, "Keith Jarrett."
14. Karen Chan Barrett et al., "Classical Creativity: A Functional Magnetic Resonance Imaging (fMRI) Investigation of Pianist and Improviser Gabriela Montero," *Neuro Imaging* 209 (April 1, 2020): 116496, doi.org/10.1016/j.neuroimage.2019.116496.
15. Charles Limb, interview with author (June 21, 2023).
16. Mihaly Csikszentmihalyi, *Flow: The Psychology of Optimal Experience* (New York: Harper Perennial, 2008).
17. "David Rothenberg on Playing Music with Whales and Nightingales," September 23, 2019, in *When We Talk About Animals*, podcast, 51:49, www.whenwetalkaboutanimals.org/2019/09/23/ep-23-david-Rothenberg/.
18. Rothenberg, *Nightingales in Berlin*.
19. Kathleen Dean Moore, *Earth's Wild Music: Celebrating and Defending the Songs of the Natural World* (Berkeley, CA: Counterpoint Press, 2022).
20. "Five Things to Know About Messiaen's Quartet for the End of Time," Carnegie Hall, March 11, 2021, www.carnegiehall.org/Explore/Articles/2021/03/11/Five-Things-to-Know-About-Messiaens-Quartet-for-the-End-of-Time.
21. Lou Fancher, "Democratic Birdsong," *East Bay Express*, July 19, 2023, eastbayexpress.com/democratic-birdsong/.

22. Lyanda Lynn Haupt, *Mozart's Starling* (New York: Little, Brown and Company, 2017).
23. Ute Eberle, "Why Soil Is a Surprisingly Noisy Place," *BBC*, February 27, 2022, www.bbc.com/future/article/20220225-the-people-eavesdropping-on-life-underground.
24. Eberle, "Soil."
25. Moore, *Earth's Wild Music.*
26. Rothenberg, *Nightingales in Berlin*, 62.
27. David G. Haskell, "When the Earth Started to Sing," *Emergence Magazine*, February 28, 2022, emergencemagazine.org/audio-story/when-the-earth-started-to-sing.
28. John Noble Wilford, "How the Whale Lost Its Legs and Returned to the Sea," *New York Times*, May 3, 1994, www.nytimes.com/1994/05/03/science/how-the-whale-lost-its-legs-and-returned-to-the-sea.html.
29. Moore, *Earth's Wild Music.*
30. Fancher, "Democratic Birdsong."
31. John Francis, *The Ragged Edge of Silence: Finding Peace in a Noisy World* (Washington, D.C.: National Geographic Society, 2011).
32. Sister Chan Khong, "Can You Hear Mother Earth?" Plum Village, July 7, 2016, plumvillage.org/articles/can-you-hear-mother-earth.

WE ARE STARDUST, WE ARE GOLDEN

1. Emmanuel Vaughan-Lee, "Finding the Mother Tree: An Interview with Suzanne Simard," *Emergence Magazine*, October 26, 2022, emergencemagazine.org/interview/finding-the-mother-tree/.
2. "Suzanne Simard, 'I say to the trees, I hope I'm helping,'" *Financial Times*, March 25, 2022, www.ft.com/content/ab6ada00-685e-499d-bd31-e975e43c5033.
3. Vaughan-Lee, "Finding the Mother Tree."
4. Krause, *Voices of the Wild*, 111.
5. Garret Keizer, *The Unwanted Sound of Everything We Want: A Book About Noise* (New York: PublicAffairs, 2010), 168.
6. Keizer, *The Unwanted Sound*, 15.
7. Dana Covit, "Why Everyone Should Try Forest Bathing," *Vogue*, March 24, 2023, www.vogue.com/article/why-everyone-should-try-forest-bathing.
8. Keizer, *Unwanted Sound*, 146.
9. David Quammen, *The Song of the Dodo: Island Biogeography in an Age of Extinctions* (New York: Scribner, 1997).
10. Weber, *Biology of Wonder*, 206.
11. Jane E. Brody, "Scientist at Work: Katy Payne; Picking Up Mammals'

Deep Notes," *The New York Times*, November 9, 1993, www.nytimes.com/1993/11/09/science/scientist-at-work-katy-payne-picking-up-mammals-deep-notes.html.

12. Krista Tippett, "Katy Payne: In the Presence of Elephants and Whales," February 1, 2007, in *On Being*, podcast, 52:31, onbeing.org/programs/katy-payne-in-the-presence-of-elephants-and-whales/.
13. Brody, "Scientist at Work."
14. Callum Roberts, *The Ocean of Life: The Fate of Man and the Sea* (New York: Penguin, 2012).
15. Moore, *Earth's Wild Music.*
16. "David Rothenberg Playing with Humpback WhalesExcerpt from French Film," David Rothenberg, February 27, 2014, YouTube video, 6:27, taken from French TV documentary, 2011, youtu.be/d-n_uZ_IYi4.
17. Carl Safina, *Becoming Wild: How Animal Cultures Raise Families, Create Beauty, and Achieve Peace* (New York: Henry Holt, 2020), 38.
18. Wikipedia, s.v. "Voyager Golden Record," last modified March 11, 2024, en.wikipedia.org/wiki/Voyager_Golden_Record.
19. Carl Sagan, *Billions and Billions: Thoughts on Life and Death at the Brink of the Millennium* (New York: Random House, 1998).
20. David Crookes, "Space Roar: The Mystery of the Loudest Sound in the Universe," Space.com, September 20, 2022, www.space.com/space-roar-loudest-sound-in-the-universe.html.
21. Sharon Heller, *Too Loud, Too Bright, Too Fast, Too Tight: What to Do If You Are Sensory Defensive in an Overstimulating World* (New York: HarperCollins, 2002), 33.
22. Ross Showalter, "Toward a Literature of Sign Language," *Literary Hub*, March 18, 2022, lithub.com/toward-a-literature-of-sign-language/.
23. Rachel Howard, "Review: William Forsythe Work Explodes Dance Prejudices at S.F. Ballet," Datebook, *San Francisco Chronicle*, February 4, 2022, datebook.sfchronicle.com/dance/review-william-forsythe-work-explodes-dance-prejudices-at-s-f-ballet.
24. Lawrence Wright, "The Elephant in the Courtroom," *New Yorker*, February 28, 2022, www.newyorker.com/magazine/2022/03/07/the-elephant-in-the-courtroom.
25. Wright, "Elephant in the Courtroom."
26. Robin Mohr, "Listening in Tongues: Being Bilingual as a Quaker Practice," QuakerSpeak, August 11, 2016, YouTube video, 5:23, comment from Virginia Jealous (@virginiajealous661), www.youtube.com/watch?v=242VqF0eYso.
27. Mohr, "Listening in Tongues."

28. Tippett, "Katy Payne."
29. Wright, "Elephant in the Courtroom."
30. Ed Shanahan, "Happy the Elephant Isn't Legally a Person, Top New York Court Rules," *New York Times*, June 14, 2022, www.nytimes.com/2022/06/14/nyregion/happy-elephant-animal-rights.html.
31. Lulu Garcia-Navarro, "Scientist Joyce Poole on What Elephants Have to Say," *Weekend Edition*, NPR, May 30, 2021, www.npr.org/2021/05/30/1001684767/scientist-joyce-poole-on-what-elephants-have-to-say.
32. "Elephant Family Project Partner Features in Late David Attenborough Film," Elephant Family, April 29, 2021, elephant-family.org/news-views/news/elephant-family-project-partner-features-in-latest-david-attenborough-film/.
33. "Conservation Partner Profile: Dulu Bora," Elephant Family, February 3, 2021, elephant-family.org/news-views/news/conservation-partner-profile-dulu-bora/ (emphasis added).
34. Garcia-Navarro, "Scientist Joyce Poole."
35. Joyce Poole, *Coming of Age with Elephants* (New York: Hyperion, 1996).
36. Wright, "Elephant in the Courtroom."
37. Krista Tippett, "Rachel Naomi Remen: How We Live with Loss," August 11, 2005, in *On Being*, podcast, 49:53, onbeing.org/programs/rachel-naomi-remen-how-we-live-with-loss/.
38. Krista Tippett, "Rachel Naomi Remen," *On Being*, August 29, 2010, onbeing.org/programs/rachel-naomi-remen-listening-generously/.

THE SOUNDS OF LOVE AND WAR

1. Nina Li Coomes, "愛してる (Aishiteru): How to Say 'I Love You' When the Language Doesn't Exist," *Catapult*, March 26, 2018, catapult.co/stories/mistranslate-column-aishiteru-how-to-say-i-love-you-when-the-language-doesnt-exist.
2. Weber, *Biology of Wonder*, 208.
3. Carolyn Wilke, "Dolphins Can Shout Underwater, but It's Never Loud Enough," *New York Times*, January 12, 2023, www.nytimes.com/2023/01/12/science/dolphins-yelling-noise.html.
4. National Research Council (US) Committee on Potential Impacts of Ambient Noise in the Ocean on Marine Mammals, "Effects of Noise on Marine Mammals," in *Ocean Noise and Marine Mammals* (Washington, D.C.: National Academies Press, 2003), www.ncbi.nlm.nih.gov/books/NBK221255/.
5. Rebecca Dunlop and Celine Frere, "Post-Whaling Shift in Mating

Tactics in Male Humpback Whales," *Communications Biology* 6 (2023): 162, dx.doi.org/10.1038/s42003-023-04509-7.

6. "Whales Give Up Singing to Fight for Love," *UQ News*, University of Queensland, February 17, 2023, www.uq.edu.au/news/article/2023/02/whales-give-singing-fight-love.
7. Carl Safina, *Becoming Wild*, 36–37.
8. Nina Li Coomes, "切ない (Setsunai): When You Need a Word to Hold Both Sorrow and Joy," *Catapult*, February 26, 2018, catapult.co/stories/mistranslate-setsunai-on-feeling-both-sorrow-and-joy.
9. Victoria Chang, *Dear Memory: Letters on Writing, Silence, and Grief* (Minneapolis: Milkweed Editions, 2021), 86.
10. Dr. Viorica Marian, phone interview with author, July 8, 2019.
11. J. Martin Daughtry, *Listening to War: Sound, Music, Trauma, and Survival in Wartime Iraq* (Oxford: Oxford University Press, 2015), cited by Alex Ross, "The Sound of Hate," *New Yorker*, July 4, 2026, www.newyorker.com/magazine/2016/07/04/when-music-is-violence.
12. Karpf, *Human Voice*, 80.
13. Krause, *Voices of the Wild*, 47.
14. Keizer, *Unwanted Sound*, 120.
15. "7 Well-Meaning Inventions That Turned Evil: 3. The Loudspeaker," Whatculture.com, whatculture.com/offbeat/7-well-meaning-inventions-that-turned-evil?page=6.
16. Roland Wittje, "Large Sound Amplification Systems in Interwar Germany: Siemens and Telefunken," Sound and Science, January 23, 2020, soundandscience.net/contributor-essays/large-sound-amplification-systems-in-interwar-germany-siemens-and-telefunken/.
17. Cornelia Epping-Jäger, "Hitler's Voice: The Loudspeaker Under National Socialism," trans. Caroline Bem, *Intermédialités* 17 (September 8, 2011): 83–104, doi.org/10.7202/1005750ar.
18. Wittje, "Large Sound Amplification."
19. Lt. Col. Dennis W. Bartow, "Combat Loudspeakers: Weapon of the Battlefield Evangelists," Psywarrior.com, www.psywarrior.com/combatloudspeakers.html.
20. Epping-Jäger, "Hitler's Voice."
21. Ilya Kaminsky, "Ilya Kaminsky on Ukrainian, Russian, and the Language of War," *Literary Hub*, February 28, 2022, lithub.com/ilya-kaminsky-on-ukrainian-russian-and-the-language-of-war/.
22. Arthur Aron et al., "The Experimental Generation of Interpersonal

Closeness: A Procedure and Some Preliminary Findings," *Personality and Social Psychology Bulletin* 23, no. 4 (April 1997): 363–77, doi .org/10.1177/0146167297234003.

23. Mandy Len Catron, "To Fall in Love with Anyone, Do This," *New York Times*, January 9, 2015, www.nytimes.com/2015/01/11/style/modern-love -to-fall-in-love-with-anyone-do-this.html.
24. David Hoon Kim, "When I Lived in French," *New York Review of Books*, July 12, 2021, www.nybooks.com/daily/2021/07/12/when-i-lived-in-french.
25. Jay Summer and John DeBanto, "Exploding Head Syndrome," Sleep Foundation, last modified December 22, 2023, www.sleepfoundation. org/parasomnias/exploding-head-syndrome.
26. Hilton Als, "The Metaphysical World of Apichatpong Weerasethakul's Movies," *New Yorker*, January 10, 2022, www.newyorker.com/magazine /2022/01/17/the-metaphysical-world-of-apichatpong-weerasethakuls-movies.
27. Eliane Brum, "Translingual," trans. Diane Whitty, *Stranger's Guide*, strangersguide.com/articles/indigenous-voices/.
28. Brum, "Translingual."

HUNGRY LISTENERS

1. *This Is Actually Happening* (podcast), www.thisisactuallyhappening.com.
2. Whit Missildine, interview with the author, Berkeley, CA, July 8, 2019.
3. Rebecca Mead, "Binge Listening," *New Yorker*, November 19, 2018, www.newyorker.com/magazine/2018/11/19/how-podcasts-became-a -seductive-and-sometimes-slippery-mode-of-storytelling.
4. Rohit Shewale, "53 Podcast Statistics: Listeners, Growth, and Trends (2023)," Demandsage, September 8, 2023.
5. Hua Hsu, "The Anti-Explainer Insight of 'Soul Music,'" *New Yorker*, December 6, 2021, www.newyorker.com/magazine/2021/12/13/the-anti -explainer-insight-of-soul-music.
6. Cited by Rothenberg in *Nightingales in Berlin*, 53–54.
7. Rivka Galchen, "The Science and Emotions of Lincoln Center's New Sound," *New Yorker*, October 10, 2022, www.newyorker.com /magazine/2022/10/17/the-science-and-emotions-of-lincoln-centers-new -sound.
8. Leonard Bernstein, "Holst: The Planets," Original CBS Television Network Broadcast, March 26, 1972, LeonardBernstein.com, leonardbernstein.com/lectures/television-scripts/young-peoples-concerts /holst-the-planets.

9. Elizabeth Howell and Adam Mann, "Why Is Pluto Not a Planet?," Space .com, October 13, 2022, www.space.com/why-pluto-is-not-a-planet.html.
10. Smith, *Talk to Me*, 50.
11. Florence Williams, *Heartbreak: A Personal and Scientific Journey* (New York: W. W. Norton, 2022), 107.
12. Florence Williams, *The Nature Fix: Why Nature Makes Us Happier, Healthier, and More Creative* (New York: W. W. Norton, 2017), 4.
13. Richard Louv, Last Child in the Woods (New York: Algonquin Books, 2008), quoted in Florence Williams, "Take Two Hours of Pine Forest and Call Me in the Morning," *Outside*, last modified June 30, 2021, www.outsideonline.com/health/wellness/take-two-hours-pine-forest-and-call-me-morning/.
14. Williams, *Nature Fix*, 5.
15. "What if they couldn't wake me up?, " *This Is Actually Happening* (podcast), February 27, 2024, Season 13 Episode 308, www.thisisactuallyhappening.com/podcast/episode/233048a5/308-what-if-they-couldnt-wake-me-up.

THE HEARING TEST

1. Jason Daley, "We Know How Stressed Whales Are Because Scientists Looked at Their Earwax," *Smithsonian Magazine*, November 16, 2018, www.smithsonianmag.com/smart-news/ear-wax-tells-story-humans-and-whales-over-last-century-180970840/.
2. Christie Wilcox, "Earwax Reveals How Humans Have Changed Whales' Lives," *National Geographic*, November 15, 2018, www.nationalgeographic.com/animals/article/whale-earwax-stress-whaling-climate-animals-news.
3. David Owen, *Volume Control: Hearing in a Deafening World* (New York: Riverhead, 2019), 240.
4. Owen, *Volume Control*, 3.

ALMOST-PERFECT RECORDINGS

1. Wikipedia, s.v. "*Selah*," last modified January 16, 2024, en.wikipedia.org/wiki/Selah.
2. E. M. Forster, *Howards End* (New York: Vintage, 1989), 31.
3. Nate Sloan and Charlie Harding, "Beethoven's 5th Symphony Is a Lesson in Finding Hope in Adversity," *Vox*, September 11, 2020, www.vox.com/switched-on-pop/21432740/beethoven-5th-symphony-deafness-switched-on-pop.
4. Sloan and Harding, "Beethoven's 5th."
5. Quoted in Williams, *Nature Fix*, 9.

ECHOES FROM THE TREE OF LIFE

1. "The Phone Booth That Allows You to Call Lost Relatives," javedJAVED javed, January 29, 2017, YouTube video, 1:58, www.youtube.com/watch?v=ZhNs_agoJB0&t=19s.
2. "After Fukushima," ARTE.tv documentary, March 11, 2021, www.youtube.com/watch?v=oixfdDLZQv8 (video has since been made private).
3. Harold M. Schulweis, "Krieh – Tearing the Cloth: Mourning," Valley Beth Shalom, www.vbs.org/worship/meet-our-clergy/rabbi-harold-schulweis/sermons/krieh-tearing-cloth.
4. En Liang Khong, "What's the Point of 'Deep Listening'?," *ArtReview*, May 30, 2022, artreview.com/whats-the-point-of-deep-listening-pauline-oliveros/.
5. "What Are the Contents of the Golden Record?" Jet Propulsion Laboratory, California Institute of Technology, voyager.jpl.nasa.gov/golden-record/whats-on-the-record/.
6. "The Golden Record, the Sounds of Earth," United Nations, www.un.org/ungifts/golden-record-sounds-earth.
7. "Frequently Asked Questions," Jet Propulsion Laboratory, California Institute of Technology, voyager.jpl.nasa.gov/frequently-asked-questions/.
8. Jamie Carter, "60 Years Ago We Saw Earth from Space for the First Time—Here's How We See It Now," *Travel + Leisure*, August 22, 2019, www.travelandleisure.com/trip-ideas/space-astronomy/history-of-earth-photographed-from-space-pale-blue-dot.
9. "Apollo 8: Christmas at the Moon," NASA, December 23, 2019, www.nasa.gov/topics/history/features/apollo_8.html.
10. Roger Payne, "I Spent My Life Saving the Whales. Now They Might Save Us," *Time*, June 5, 2023, time.com/6284884/whale-scientist-last-please-save-the-species/.
11. Elizabeth Kolbert, "The Strange and Secret Ways That Animals Perceive the World," *New Yorker*, June 6, 2022, www.newyorker.com/magazine/2022/06/13/the-strange-and-secret-ways-that-animals-perceive-the-world-ed-yong-immense-world-tom-mustill-how-to-speak-whale.
12. Wikipedia, s.v. "Charles Limb," last modified March 27, 2023, en.wikipedia.org/wiki/Charles_Limb.
13. Yao Xiao, "I Wanted to Write and Teach Literature Like My Mother Did," *Electric Literature*, June 21, 2023, electricliterature.com/i-wanted-to-write-and-teach-literature-like-my-mother-did/.

ELIZABETH ROSNER is a bestselling novelist, poet, and essayist. Her works include *Survivor Café: The Legacy of Trauma and the Labyrinth of Memory*, a finalist for the National Jewish Book Award, and the novel *Electric City*, named a best book by NPR. Rosner's essays have appeared in *The New York Times Magazine*, *Elle*, and numerous anthologies. She lives in Berkeley, California. Find out more at elizabethrosner.com.